10/26/05

A wealth of infor
but only theories abor
America got its name

Wm. J Kres Sr.

THE True Story OF How America Got Its Name

RODNEY BROOME

Truth is the daughter, not of authority, but of time.
—Francis Bacon

MJF BOOKS
NEW YORK

To my grandfather, Captain William Broome
and my father, Cecil Broome

Published by MJF Books
Fine Communications
322 Eighth Avenue
New York, NY 10001

The True Story of How America Got Its Name
LC Control Number 2002106260
ISBN 1-56731-545-3

Previously published as *Terra Incognita: The True Story of How America
Got Its Name*

This edition published by arrangement with Educare Press

Manufactured in the United States of America on acid-free paper ∞

MJF Books and the MJF colophon are trademarks of Fine Creative Media,
Inc.

BG   10 9 8 7 6 5 4 3 2 1

Table of Contents

Illustrations and Maps

Acknowledgements

A writing opportunity like this comes along but once in a lifetime, and I am grateful for the opportunity afforded to me. I would like to thank all the people who have helped me assemble the information that was needed to write this book.

I particularly want to thank my cousin Peter Martin who has lived his life in Bristol and is a local history enthusiast. He was expressly involved in the John Cabot quincentennial celebrations that took place in Bristol in 1997. At that time, the Matthew Project built a replica of the *Matthew* and sailed it across the Atlantic to Canada and back, replicating the original route.

I was aware of the story of Richard Amerike, as is everyone from Bristol, but I was not familiar with the details. Peter did in-depth research and was working together with Peter Macdonald on a publication about John Cabot, Richard Amerike, and other local families of fifteenth-century Bristol.

On a trip to Bristol after I had begun writing this work, I met Anna Hurl, who had spent countless hours in the Bristol records offices and libraries researching and accumulating material for Peter Martin. I obtained many of the files and copies of manuscripts she had collected, and these provided much in-depth information, a fact that saved me hours of research that I could not have achieved during my short vacation there. My thanks to Anna.

Much later, I viewed a film that was produced by the Bristol Film and Video Society entitled "John Cabot–The Story." This wonderful production brought the account of Richard Amerike to life for me.

My thanks also to Lesley Nichols, Dave Arnold, Ken Squire, John Spiller, Malcolm McDowall, Andy Nash, John Waldren, Dr. Norman Morgan, and, last but not least, to my mother for providing informational documents and for doing some last-minute running around on my behalf in Bristol and London.

Bristol's resident history expert, Anton Bantock, is acknowledged in almost all the local books I referenced, and I too want to thank him for contributing so much on the subject. He has graciously allowed me to reproduce several of his sketches. Anton is the founder and vice chancellor of the University of Withywood, an educational institution and charity that is dedicated to furthering studies in both the local area in Bristol and developing nations.

Many thanks to Mr. Hubert Williams and his daughter Wendy for an illuminating tour of St. Mary Redcliffe Church. Mr. Williams is just one of several ex-Mayors of Bristol named in this book.

Also I am indebted to the Master of the *Matthew*, Nigel Ottley, and crew members Shawn and Jean for a below-deck visit. The present owners of Richard Amerike's Ashton-Phillips house, Donna and Steven Brown, gave me an enthusiastic tour of the house and copies of several important documents, in particular the newspaper article from the Western Daily Press.

The Reverend Beverly Tasker, vicar of All Saints Church, Long Ashton, which is adjacent to Amerike's property and is believed to be the burial site of some of his descendants, enlightened me with the history of the village.

Thanks to Gerry Brooke of Bristol United Press for permission to use the newspaper article from The Western Daily Press.

I wrote Terra Incognita in Seattle, where I have lived for thirty years. A special thanks to Lorraine for her help and understanding. She is responsible in part for the photograph on the jacket cover.

Thanks also to Karl Pelkan for some very useful leads and to Lars Haaheim and Martin Lode for imparting their firsthand knowledge of the salt cod and stockfish trade. Thanks to Kieran O'Mahony for his extensive help in making this book a reality. Finally, thanks my publicist, Maryglenn McCombs, to my editor, Marianne Auwaerter Van de Vrede, and to Jeff Reynolds, the book's designer.

Preface

"The fairest, goodliest and most famous Parish church in England."

QUEEN ELIZABETH I, AUGUST 1574

St. Mary Redcliffe Church sits atop a red cliff on the bank of the River Avon in the English city of Bristol. It was the merchants' church, and 500 years ago it had a congregation of well over 100 affluent merchant traders and their families. Fortunes were being made trading in wine, wool, leather goods, and fish.
The ships belonging to these merchants were moored in the river and sailed to foreign ports from Africa to Iceland. Bristol was an international port.

We stood on the gravel path outside St. Mary Redcliffe Church, squinting skywards at the towering spire. The pointed edifice cast a ragged shadow across the river and the quays toward Queen Square and the neighborhood where my father lived when he was young. I didn't know it then, but it would cast an even greater shadow to the west, past the narrow streets, past the spilling tidewater, and across the Atlantic to a new world that would one day be home to me.

We had been walking at a fast pace through the narrow streets and along the quays of the old city one Sunday afternoon. I can still hear how my father's shoes clicked rhythmically on the cobblestones in the nearly deserted streets. It was late in the autumn of 1952. I was eight years old at the time. My dad had grown up in these streets, and when he looked at the closed businesses and shuttered warehouses, it was with an air of quiet expectation of the activity and commerce that would soon signal the start of another week.

It was foggy, damp, and piercing cold, and the wind whistled up from the river. The "North Pole" was what he always called the corner where the River Frome meets the River Avon. And that is the name by which I have referred to it ever after.

In the church, he showed me the giant whalebone standing in the small chapel just inside the main doors. I recall it being about twice my height, and we added our fingerprints to the shiny, black surface. My dad told me that John Cabot brought this back with him from Newfoundland 500 years before.

Growing up as a lad in Bristol meant living the legends of the great discoverers and many other colorful characters of the sea. Edward Teach, the notorious pirate known as "Blackbeard" grew up in these streets, as did Captain Woodes Rogers, the Governor of the Bahamas, who was responsible for Blackbeard's death in a sea battle off the North Carolina coast in 1718. The same Captain Woodes Rogers rescued Alexander Selkirk, a seaman stranded on a Pacific island. Daniel Defoe met them in a waterfront public house in Bristol, and his classic novel, *Robinson Crusoe*, was the result. The pirates William Dampier, a buccaneer who later became a famous scientist, and Henry Morgan were also familiar figures in Bristol streets. Even Robert Louis Stevenson's fictional "Long John Silver" reputedly frequented two of Bristol's public houses that exist to this day.

Merchant ships came right into the center of the city as they had done for hundreds of years. It was a hive of mercantile activity where fascinating deliveries happened daily and business seemed to overflow into the numerous pubs. The brewery, situated on the river next to Bristol Bridge, still used horses to draw the beer drays, and the barrels were rolled off noisily and steered down the alleys into the cellars. The wooden cellar hatches opened up into the narrow pavements outside the pubs.

Sitting on the quayside were casks and barrels of sherry, port, and wine that came off the ships. The smells of the city changed between the street and the alley, for at the end of every street was the river. The ships unloading their cargoes brought an assortment of sweet and acrid aromas that mixed with the everyday river offering. Fish arrived fresh and bloody in boxes; hops were heavy with fragrance for the brewery; sherry and wine came in foreign casks with romantic Mediterranean names; bananas and rotting food caused the seagulls to make a ruckus. Sawdust was sprinkled liberally everywhere so that the spillages could be cleaned up daily. Added to these were the industrial smells of timber, coal, and coal gas, which was manufactured in the works behind the cathedral.

My grandfather was not a native Bristolian. He was born in 1860 and raised in Conwy, or Conway, a small town on the North Wales coast. As a schoolboy, he would look out to sea at the sailing ships leaving from Liverpool, some sixty miles to the east along the coast. Many of these ships were sailing to America. The sea was in his blood, and within a few years, he moved to South Wales and then to Bristol. In December 1899, he gained his Master Mariners certificate, and he entered the new century as the captain of his first ship.

He sailed on a merchant sailing ship called the *Morven* for several years between 1896 and 1900. He captained two other merchant ships, the *Menantic* and the *Mohican*, during the first decade of the twentieth century, sailing between Bristol and the United States. After the outbreak of World War I, he delivered petroleum products to France. I still imagine him crossing the Atlantic under sail to Canada, hauling Welsh coal outbound and Canadian wheat back to Bristol. He was a member of the Society of Merchant Venturers, the organization that took care of my grandmother and the children's education after he died in 1917. My dad spoke very highly of it.

Their house was directly on the quay, where the merchant ships moored, and they were right across the river from St. Mary Redcliffe Church, reputed to be the most magnificent parish church in England. The spire of the church had only recently been rebuilt, having been destroyed by a bolt of lightning 400 years earlier.

My father and his brother both attended St. Nicholas Church primary school, which was just a quarter-mile walk from the house and across the street from Bristol Bridge. St. Nicholas Church is the sailors' church.

As children growing up in Bristol, we knew about John Cabot, whose statue is in front of the civic building. Likewise, we had heard of the *Matthew*, the ship that took him to North America. My dad also relayed stories of the Bristol fish traders who went as far as Newfoundland years earlier.

Bristol, with its westward perspective, has always had strong connections with America. Sixty percent of all trade passing through Bristol was destined for or from the American colonies before the Revolution in 1776. The first United States Consulate was opened there in 1792, even before the country was recognized with an embassy in London. The wheels of trade were more important than political niceties.

In 1897, Bristol was commemorating the 400th anniversary of John Cabot's voyage in the *Matthew*, when he supposedly became the first European to set foot on American soil. The focus and the festival preparation caused the unearthing of many new bits of evidence that when pieced together bore semblance of a tantalizing story surrounding the events that took place toward the end of the 1400s. The discovery of a letter in Spain in 1955 and shipping records in London in the 1960s have considerably strengthened the evidence supporting what had long been suspected regarding the discovery and naming of America.

A hundred years later, in 1997, the city of Bristol reconstructed a replica of the *Matthew* and relived Cabot's voyage to Newfoundland.

Three years ago when I first began working on this story, I gave it no thought that I would also be retracing my grandfather's journey from the quay in Bristol across the Mare Oceanum to America. I had no idea that this journey through time would lead me to my own roots in North Wales.

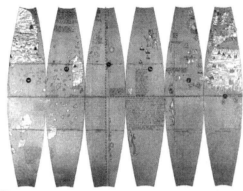

Introduction 🖋

Terra Incognita, the fourth continent, lay dormant as if shrouded in a fog until the year 1502. The native peoples of the continent– the Inuit, Micmac, Aztec, Inca, and Apache–had all built sophisticated societies, but they had not participated in the ambitions of a Euro-centric world that had expanded to encompass Africa and Asia.

Visitors had been to its shores before. Ships had arrived, and the crews did what they had gone there to do. They cut trees, they fished, they settled for a season, and they left.

What was different about Christopher Columbus's arrival at the Caribbean Islands in 1492, and, five years later, John Cabot's voyage to New England and Nova Scotia was that neither man was interested in the land he had reached. They were both trying to get to Asia.

Naturally, they were frustrated because they could not get around the vast landmass that blocked their way. Columbus was trying to reach India and Cabot, China, and in fact both men thought they had reached their destinations, taking that belief with them to their graves. It was the political opportunism of others that elevated their voyages into "discoveries."

Earlier visits had not occurred at the time of a rapidly expanding international economy. According to legend, an Irish monk arrived there over a thousand years ago. Later, Icelandic and Norwegian wayfarers visited in the tenth century and stayed for 300 years. Portuguese sailors were reputed to have visited the Caribbean in 1424 and again in the 1470s. Basques reached Newfoundland in the late 1400s, a Dane claimed to have discovered it in 1472, and merchant fishermen from England visited many times from approximately 1480 onward.

Most of the explorers involved in the discovery of the Americas in the late 1400s either knew or were known to each other. The paths of many of these colorful characters crossed each other, sometimes more than once.

The only person who did not meet any of the central characters was a mapmaker whose last name he invented from his geographical surroundings: Martin Waldseemüller. Ironically, the confluence of a series of historical coincidences would distort and influence our world to a degree out of all proportion to their significance. Waldseemüller's fluid pen and the timely invention of movable type had a lasting effect on the way we view our world today.

THE True Story
OF How America
Got Its Name

1.

Twelve Wooden Plates

*"It is well here to consider the injury and injustice which
that Amerigo Vespucci appears to have done to the
Admiral (Cabot),...in attributing the discovery of this
continent to himself...Owing to this, all of the foreigners
who write of these Indies in Latin,...call the continent
America, as having been first discovered by Amerigo.
For as Americo was a Latinist, and eloquent, he knew
how to make use of the first voyage he undertook, and
to give credit to himself, as if he had been the principal
captain of it."*

FATHER BARTHOLOMEO DE LAS CASAS,
BISHOP OF CHIAPAZ, 1559

Martin Waldseemüller published a revolutionary World Map
in 1507, only fifteen years after Columbus landed on the
Caribbean island of Hispaniola. It was the first map to show
the Americas as a separate continent and accurately depict the length of the

1

equator. On this map the word "America" was written across the part of the continent we now call Brazil. It was produced and printed by a group of men working in a Benedictine monastery in St. Die, a small town located in the Vosges Mountains near Strasbourg in Lorraine. Lorraine was a separate principality located between Germany and France, and was not ceded to France until the Treaty of Chambard in 1552.

In 1505, Rene II, the Duke of Lorraine, retained Waldseemüller to oversee the production of the map. Waldseemuller was thirty-four years old at that time, an accomplished cosmographer and a practicing cartographer. The Duke and the Canon of St. Die had formed an intellectual group known as the Gymnasium Vosgense in 1500 to expand the rudiments of cosmography and geometry. Waldseemüller joined Gautier Lud, secretary to the Duke, his nephew Nicolas Lud, Jean Basin de Sandancourt, and Matthias Ringmann, a poet and teacher of Latin and Greek. Gautier Lud was a wealthy man who owned one of the newly invented printing presses. It is no surprise that Waldseemuller, Ringmann, and other cartographers were associated with the church of St. Die since the clergy made up the majority of the lettered class and were the chief patrons of the book trade in the years leading up to the Renaissance and long after.

Waldseemüller was born in the village of Wolfenweiler, near Freiburg, a small town in present-day Germany close to both the French and Swiss borders, sometime around 1471. In 1480, the family moved to Freiburg, where his father became a member of the city council in 1490. The same year Waldseemüller became a student at Freiburg University. He had a lively interest in the imagery of names. Early in his career, he reinvented himself as Hylacomylus in an attempt to elevate his stature. Hylacomylus was a stylized translation of his name from the Greek for wood (wald), the Latin lacus, or lake (see), and the Greek for mill (müller). His accomplice, Ringmann, similarly reinvented himself as Philesius Vogesigena. Philesius is simply a Latin pen-name that occurs in classical texts. Ringmann, like other humanists, assumed a classical pen name that incorporated the Latinized place reference "Vogesigena" obviously bearing reference to St. Die.

✠

It was well known that the Spanish and Portuguese were exploring to the south along the West African coast and across the Mare Oceanum, but the authorities were secretive about their findings and only released details when they were ready to make land claims. It was 1503 before they released a serious amount of information, which was eagerly studied by the cosmographers and cartographers. By this time, Christopher Columbus was on his fourth expedition to the Caribbean area. The first was in 1492, and the second was soon after his return, in the following year. During his third voyage, from 1498-1501, he established a settlement on the island of Hispaniola.

Amerigo Vespucci was another Italian navigator who made several voyages and participated in Columbus's third expedition. He was an educated man who wrote and published books about his voyages. He also sent long detailed letters to Peter Soderini, a boyhood friend in Florence. These letters were also published separately.

In 1505, a Latin edition of Vespucci's "third voyage" was printed in Strasbourg and was illustrated with wood-cut printing plates. Matthias Ringmann wrote the dedication and a poem. Amerigo Vespucci was evidently aware of the work of the Gymnasium Vosgense, and he may have had frequent communication with Ringmann. Ringmann and Waldseemüller obtained the new edition of *The Four Voyages of Amerigo Vespucci*, a published collection of scientific letters and manuscripts he had written to Soderini, and the accompanying maps. Vespucci may have sent them himself.

Vespucci planned to produce a World Map based on his research, and some think that he communicated this to Ringmann and Waldseemüller. He recorded that he planned to make two world maps, one a flat page and the other a globe. This is exactly what Waldseemüller did. It is likely that Vespucci had drawn rudimentary maps and Waldseemüller elaborated upon them.

The Duke Rene and Vespucci may have known each other. The Duke had a keen interest in cosmology and visited Toscanelli in Florence in the 1470s while Vespucci was still studying physics and cosmography there as a youth. Vespucci moved to Paris in 1478, where he was attached to the Florentine Embassy. The Duke was known to have visited there.

Strasbourg was the birthplace of the printing industry, and Gutenberg set up one of the first printing presses there in 1475. Gutenberg's invention of movable type set in motion a revolution in intellectual opportunism. The Duke of Lorraine and Lud saw the commercial possibilities in producing the most current World Map, and the impact of this would prove far-reaching in its implications.

Waldseemüller, who was known for his work on maps and his woodcut publishing abilities, started work on the map. He established the length of the equator from the calculation as determined by Vespucci, and he relied on Ptolemy's Geographia as the basis for Europe, Africa, and Asia.

For his depiction of the eastern coastline of the New World, he faithfully reproduced a map published in 1505 by an Italian mapmaker named Nicolo Caveri. In 1502, Alberto Cantino, an Italian diplomat in Lisbon, illegally obtained a Portuguese map of the New World, which he smuggled to his employer, the Duke of Ferrara, Ercole d'Este. Caveri's map duplicated the Cantino map and added some coastline in the Gulf of Mexico. These maps were sent from Italy to Waldseemüller's group by the Medici family.

The remainder of Waldseemüller's map of the New World relied entirely on Vespucci's manuscripts, *Mundus Novus* and *The Four Voyages of Amerigo Vespucci*.

Waldseemüller's group took two years to prepare the map. It was carved into twelve large wooden plates for printing. Each plate was 21" x 30". When assembled, the map measured approximately 8' by 4.5', about the size of a blackboard. This map, the first to show the New World in its entirety, is a bay window into the ancient mysterious. Here are the precisely limned lands of Europe. Africa's coasts are labeled in great orderly detail, and its interior is free, clear, unnamed, and unexplored. It originally showed North and South America as two separate continents separated by a narrow gap. The name "America" is used to designate part of the feathery landmass we now know as South America. The general shape of South America is surprisingly accurate, and the west coast of North America is portrayed as mountainous.

As the map was nearing completion in 1507, information came from Columbus's fourth voyage that he was unable to find the passage to India or the Pacific Ocean. Waldseemüller added a small inset at the top of the map, correctly showing the isthmus between North and South America with no gap. Columbus could not have known of the narrow isthmus from his own travels.

✠

Most historians agree that the name "America" was written on at least one of Vespucci's maps, but Vespucci did not name the continent or claim that it was named after him. In his publication, he called the new land "Quarta Orbis Pars" (the Fourth Part of the Earth).

Many scholars believe that Vespucci's maps included the "secret map" of the lost mariners that Columbus is reputed to have had in his possession in 1492. While charting the coast from Venezuela to Mexico, these lost mariners reached the part of Central America that is now Panama. It is believed that natives gave them information about a canoe passage to another ocean there. This gave rise to the belief that there was a "gap" through to India. Waldseemüller had extensive information that there was another large ocean on the other side of a narrow isthmus.

The Portuguese authorities had reprimanded Vespucci for releasing a map they considered to be a state secret.[1] Columbus's secret map might have been the document in question.

✠

In 1497, another Italian navigator, John Cabot, sailed in an English ship from Bristol and is credited with being the European who discovered the North American Continent. This expedition was not yet widely publicized and the maps and records were not published.

In 1498, Cabot and several English ships ventured again to North America and ultimately fell into conflict with the Spanish explorers. Vespucci got to see Cabot's maps and records of his voyages.

Because Vespucci had seen the Cabot maps, Waldseemüller had the benefit of the Cabot survey of the Eastern Seaboard of the New World without knowing who had charted it. Vespucci would have edited any locations named by the English that had been written on Cabot's charts when he transferred them to his maps.

✠

There was considerably more geographical knowledge in circulation than was learned from the Columbus expeditions. How could Waldseemüller have known that North and South America are two separate continents connected by a narrow isthmus and that the west coast of North America is mountainous? Vespucci probably obtained information regarding the west coast from native peoples who were able to describe the narrow isthmus and the vast ocean on the other side.

It would be another six years before de Balboa would cross the isthmus and be the first European to see the Pacific Ocean, and another twenty years before Magellan would sail from South America to Asia and confirm Waldseemüller's overall view of the world.

Waldseemüller was unable to account for the origin of the name America, but he added text to the map suggesting that it could perhaps have been derived from the first name of the Italian navigator Amerigo Vespucci. The finished map portrayed likenesses of both Ptolemy and Vespucci in honor of these two great world figures.

The following translation is part of the Cosmographiae in which the mapmakers attempted to justify how the name America must have been derived from Amerigo Vespucci:

"But now these parts (Europe, Africa, and Asia) have been extensively explored, and a fourth part has been discovered by Americus Vespuccius; I do not see what right any one would have to object to calling this part Americus; who discovered it and who is a man of intelligence, and so to name it Amerige that is the land of Americus, or America, since both Europa and Asia got their names from women."

Waldseemüller and his group evidently believed that Vespucci discovered South America. Columbus was given credit for the discovery of the Caribbean, but there is no mention of Cabot and the English voyages. Matthias Ringmann may have put forward the theory of the origin of the name America, since it was he who wrote this text. Ringmann was only twenty-one years old when the project started; he was young and probably in awe of Vespucci.

Waldseemüller's spelling of the name America strongly supports the argument that he saw it in writing. Spain did not recognize the name for 300 years, but it is believed that in England it was independently gaining widespread use.

"…the fourth part of the world, which more commonly than properly is called America."

RICHARD HAKLUYT, ENGLISH AUTHOR OF
PRINCIPALL NAVIGATIONS, 1589

By 1538, its use was so prevalent that Mercator used it on his map for both North and South America.

Waldseemüller's World Map was published on April 27, 1507, and over 1,000 prints of the first edition were distributed. It is likely that by 1508 every university, library, and government map room in Europe had a copy. It was well known to historians despite the fact that all the known copies had disappeared by the eighteenth century.

The map was unique in more ways than one. It was the first to show the world with the new fourth continent. This new concept of the world was a fascination to everyone–politicians, merchants, and adventurers alike. Europeans from several nations were now looking to explore and exploit this new land for the wealth it promised.

The map was a mass-produced product of the new printing press, and its widespread propagation throughout Europe was one of the first consequences of typography. The printing press was embraced by scholars, artists, and businessmen alike to further education, industry, and art, not unlike today's World Wide Web. Up to this time, all maps had been individually drawn. They were expensive, few in number, and not widely seen. This map was being prominently displayed in hundreds of institutions that had not had a map of any significance up to that time. Thousands of people, many of whom had never seen a map before, saw and studied Waldseemüller's map.

✠

"He (Amerigo Vespucci) is said to have placed the name America in maps, thus sinfully failing towards the Admiral."

BARTHOLOMEO DE LAS CASAS, 1559

Father Bartholomeo de Las Casas, Bishop of Chiapaz, was not the only person to protest the prominence given to Vespucci. Father de Las Casas was writing a biography of Columbus. A short time after the distribution of the first edition of the 1507 World Map, the Duke of Lorraine and the group at St. Die were contacted. There were protests about the prominence of Vespucci's contribution to the map and the almost total exclusion of Columbus's role. Waldseemüller made an attempt to reverse his error. Later editions of the World Map dropped the name America and all references to Vespucci.

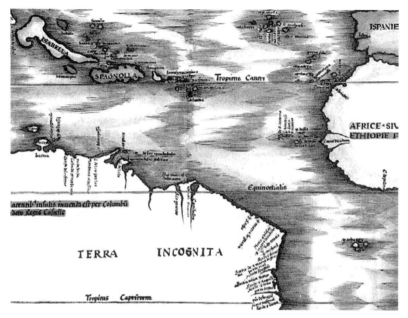

MARTIN WALSEEMÜLLER'S MAP OF 1513. THE WORD
"AMERICA" FROM THE 1507 EDITION IS CHANGED TO
"TERRA INCOGNITA." CREDIT FOR THE DISCOVERY OF
AMERICA IS ALSO CONFERRED ON COLUMBUS.

In 1513, Waldseemüller published a seafaring chart of the Caribbean
area, including a portion of South America. The name America did not
appear at all, and the southern continent that he had previously called
America is merely designated as "Terra Incognita," an Unknown Territory.
He removed all of Vespucci's contributions to the 1507 map: all indications
of a separate continent, the isthmus, the west coasts of the Americas,
the name America, and any reference to Vespucci. Finally, and most
significantly, he added a paragraph in which he gave Columbus credit for
discovering the New World.

However, it proved impossible to erase the effect of the widespread
distribution of the first edition. After its publication, it remained the
primary reference map for over twenty years and was copied by several
other mapmakers. Peter Apian used the name America in exactly the same
way in his 1520 World Map.

9

Waldseemüller didn't know the origin of the name America, but it clearly predated the period when he drafted the map. The name America almost certainly was in common use in England in the early 1500s by the English seafaring community, who were regularly engaged in steering a course towards the fishing grounds of Brassyle. However, the concept of this land being a separate continent was not realized in Bristol until sometime after 1507, perhaps when a copy of Waldseemüller's map arrived in the city. Bristol ships were still searching for a northwest passage through the islands to Cathay.

By the 1530s, it is likely that the southern part of the New World was named America solely due to Waldseemüller's map and the many maps published by others who copied it. It is possible, in any case, that the northern half would have been named America because of its continuous use by English mariners and merchants for over forty years.

In 1538, Gerard Mercator published his World Map in the central European city of Duisberg, in the Duchy of Cleves. It replaced Waldseemüller's map as the authoritative view of the earth. He extended the use of the name America by writing "North America" and "South America" on the two parts of the continent, and it has been that way ever since.

Whether or not Waldseemüller can be accused of recklessness for giving credit to Vespucci is moot. The map at this pivotal point in history that bade a nebulous goodbye to Ptolemy, heralded Vespucci, and ignored Columbus and Cabot was thrust into the forefront of academe by means of a modern invention–the printing press.

✠

By the 1830s, the name America was fully established, but without any copies of Waldseemüller's map in existence, the derivation of the word remained a mystery. The theory that America originated from Amerigo Vespucci's name dates from 1838 when some documents referring to the Waldseemüller map were discovered in a Paris market.

The intriguing new theory that the name enjoyed acceptance in England as early as 1500 surfaced around 1900.

✠

Waldseemuller's map surfaced in 1901 at the family castle of Prince Johannes Waldburg-Wolfegg in southern Germany. The Library of Congress and others made some efforts to acquire the map. Some 500 copies of the map were made, and the Library owns one reproduction.

Prince Waldburg-Wolfegg has offered to sell the map to the Library of Congress for $10 million. An extra $4 million would buy the rare sea chart known as the Carta Marina, also drawn by Waldseemuller, and two incomplete sets of celestial gores. The Carta Marina was based on unpublished nautical charts kept secret by explorers. It is the first printed nautical chart of the modern world. The borders are decorated, and the map is alive with cartouches, festoons, and ornate illustrations. It includes a drawing of an exotic opossum in South America and a rhinoceros in Africa.

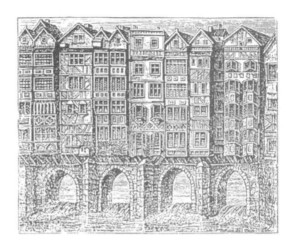

2.

The Commercial Revolution

"Spices grow in the lands to the west,
even though we usually say to the east,
for he who sails west will always find
these lands to the west,
he who travels to the east by land
will always find the same lands in the east."

PAOLO DAL POZZO TOSCANELLI (FLORENCE),
JUNE 25, 1474, LETTER TO CANON FERNAO MARTINEZ,
CANON OF LISBON CATHEDRAL

The Black Death, the ratborne plague that decimated Europe between 1346 and the early 1400s, exacted a terrible toll on the population. One third of the people died, and an economic stagnation that lasted for 100 years followed. However, by the mid 1400s, the populations of European countries were rapidly increasing, and an affluent merchant class was emerging. This was the beginning of the "Commercial Revolution."

The urban centers in Europe changed dramatically during this time. Paris was the largest city in central Europe with a population of nearly 100,000. It dominated a rural country scattered with small urban centers.

Almost as large and 250 miles to the north and across the English Channel was London, the capital of England. England was also largely an agricultural nation with many small market and county towns. Paris and London have always been rivals, and the 1400s were a particularly adversarial time.

The third major city in central Europe was the port of Antwerp. It is situated on the Schelde River, adjacent to the delta of the River Rhine, the major artery into the heart of Europe. Along this river, virtually all the export trade flowed from Germany, eastern France, Switzerland, and Belgium.

Many English merchant ships plied the waters trading between London and the ports of Calais, Antwerp, Amsterdam, and Bergen, though frequent political problems and changes in customs duties affected trade. London dominated foreign trade in England, with sixty percent of the goods leaving and entering England passing through its port.

In Europe, there was a tremendous and growing demand for spices, silk, and other exotic goods from the East. In 1271, a young Venetian by the name of Marco Polo set out with his father and uncle on a journey through Asia and as far as Cathay. During these travels, Polo marveled at the many spices that were used both for cooking and medicinal purposes. He wrote of Java, "from thence also is obtained the greatest part of the spices that are distributed throughout the world." Polo wrote that the Indian city of Delhi "produced large quantities of pepper and ginger, with many other articles of spicery."

There were established overland trade routes,[1] and he wrote of the city of Ormus, situated at the entrance to the Persian Gulf, that it was the "port frequented by traders from all parts of India who bring spices and drugs… These they dispose to a different set of traders, by whom they are dispersed

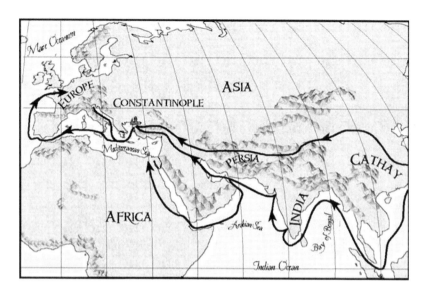

PRINCIPAL SPICE TRADE ROUTES

throughout the world." On his return to Venice, he brought some of the exotic spices and silk goods that he had accumulated during his travels.

From the foundations laid by Marco Polo, the Venetians had monopolized the spice trade by 1450. These goods traveled overland along the spice roads through Persia to Asia Minor, or by sea from India and Cathay to the Arabian ports and then overland, the routes converging in Constantinople. There, Venetian merchants loaded the spices onto their ships, and they were carried to Venice and other Mediterranean ports for distribution into the European market. Just once a year, a convoy of Venetian galleons carried spices through the Mediterranean, out into the Mare Oceanum, and north to England, Holland, and the Baltic region. The profits made by the Venetians were staggering. It was said that if they lost five out of six ships they would still show a profit.

Pepper and ginger from India were particularly in demand. Pepper, the more valuable of the two, was used in cooking and also as a tonic, a stimulant, an insect repellant, and an aphrodisiac. Cinnamon came from Cathay and Burma and was used for cosmetics, drugs, balms, oils, and perfume, in addition to flavoring food. Nutmeg came from the Banda Islands and cloves from the Moluccas, also known as the Spice Islands.

The affluent used spices extensively for cooking. After the autumn harvest, there was very little fresh food available in Europe until the following spring. In the northern countries, food was not very palatable, particularly in winter. Spices, however, especially pepper, enhanced the taste of meat and other foods. It became a mark of society to have these exotic tastes in the food one served, even when the meat was fresh.

Throughout the fifteenth century, the demand for spices rapidly increased, buoyed by burgeoning populations and spreading affluence.

The supply of spices was suddenly disrupted in 1453 when the Turks took control of Constantinople. The Christian Byzantine Empire was defeated, and the rulers were driven back into the Balkans. The spice route was blocked, and the traffic was diverted south through Alexandria in Egypt. The goods got through to the markets, but transportation costs greatly increased and the countries en route exacted a high toll in customs duties. The price of Indian pepper increased thirtyfold during the last half of the fifteenth century.

The economic incentives offered by the spice trade encouraged Portuguese and Spanish merchants to step up exploration efforts to find a different route to bring the spices to Europe. They would then be able to compete with the Venetians. The Portuguese, in particular, initially concentrated their efforts in attempting to sail around Africa and reach India by sea. Ships would be able to sail from Asia to Western Europe without having to unload their goods and transport them overland. Merchants could import them unimpeded and duty free.

In 1475, an Italian cosmologist by the name of Toscanelli produced a map that offered a totally new concept. He showed that it would be possible to reach India by sailing west across the Mare Oceanum. The ports facing that sea on the western edge of Europe would now have a new significance, one that would not be lost on Christopher Columbus.

3

A Young Genoan Arrives in Bristol

*"To this island (Iceland), which is as big as England,
the English merchants go, especially those from Bristol…
there were vast tides, so great that they rose and fell as
much as twenty-six fathoms in depth."*

JOURNALS OF CHRISTOFORO COLUMBUS[1]

In October 1476, Christoforo, a 26-year-old Genoan, was working as
an ordinary seaman on a Genoan merchant ship sailing between the
English ports of Southampton and Bristol. The ship had rounded the
Cornwall and Devon peninsula and was in the wide estuary, known as the
Bristol Channel, that separates Wales and England. Along with several
other vessels, they were approaching the River Avon. Two or three hours
after the tide turned from low to incoming, the skipper sailed the vessel
into the River Avon for the 6-mile trip up the river to the city of Bristol.

Four rowboats, each with six oarsmen and a coxswain, rowed out from the village hamlet of Pill to meet the ship at the mouth of the river. Ropes were thrown down and attached to the rowboats. The rowboats then towed the large ship, guiding it in the river. The swelling tide carried the ships at a speed faster than a man can walk.

The ships headed up the river in a convoy. After they had been traveling for about 4 miles, they found themselves in a narrow gorge flanked by cliffs that were close to vertical and towered 300 feet over the river. They traveled past the Manor of Clifton and the Abbey of St. Augustine and then sailed into the very heart of the city of Bristol, which was a bustling community of 10,000 people, arriving just before high tide.

The ship was tied up at the quay, and the crew prepared to unload the cargo. At high tide, Christoforo noticed that ships were getting ready to leave on the outgoing tide. The rowboats that had piloted the Genoan ship into the center of the city were then assigned to an outgoing vessel. Departing ships were carried along by the current, back through the gorge to the mouth of the river, and out into the Bristol Channel, where the full sails would be raised.

Six hours later, the ships that had remained in the harbor lay stranded on the muddy river bottom. The river had dropped to little more than a stream just a few feet wide. This phenomenon repeated itself twice each day, and it was something Christoforo had never witnessed before. He was astonished by tides such as these, where the sea fell a good thirty feet in just six hours.[2]

Christoforo and the crew of his ship soon found the taverns where the Genoan community congregated. He used this opportunity of being in Bristol to find work as an ordinary seaman on a merchant ship going north to the fishing grounds of Iceland. He would have been steered to the Church of St. Nicholas, the patron saint of sailors, and to the Mariner's Institute on Marsh Street. There he would have found out when the ships were sailing and who was hiring.

Two or three large merchant ships sailed from Bristol to Iceland every year, and Christoforo was hired to work on a vessel that sailed in February

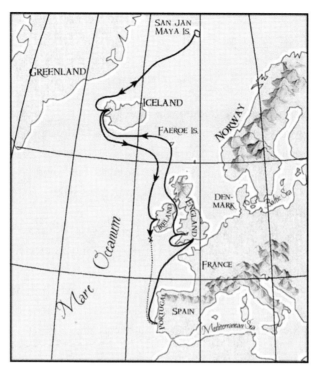

CHRISTOFORO COLUMBUS'S TRAVELS IN 1476 AND 1477

1477. The ship may have belonged to the Canynges Company, which owned the largest fleet of ships in Bristol and legally controlled all fish trade with Iceland.[3] In the 1450s and 1460s, the Icelandic dried fish trade contributed handsomely to the Canynges Company's profits. The founder, William Canynges, who had died two years before, had been both Mayor and Member of Parliament for Bristol.

The ship's cargo contained a variety of utilitarian goods, mostly produced in the west of England. Cheese, butter, honey, and grain were always in demand, and the Bristol ships also supplied metal tools, cheap Flemish linen, and basic commodities, such as vinegar, salt, and wood. Christoforo would have noticed the trademarks[4] chalked on the lids of the barrels, identifying the merchants who owned the contents. Typically, three or four merchants would have had cargo on board. The accounts of each merchant were kept separately.

✠

The ship sailed the traditional route: out into the Bristol Channel, heading west along the South Wales coast, and then north into the Irish Sea, following first the Welsh coastline and then the west coast of Scotland. Sea crossings to Ireland were extremely rough, and a high level of seaworthiness was required of the vessels undertaking these voyages to the edge of the Mare Oceanum and beyond.

Trade with Iceland had grown out of the regular Irish trade as early as the 1420s, and Galway had become a primary staging port for many of the ships, in particular the smaller ones.

They were now out in the open ocean and sailed north across 500 miles of the north Mare Oceanum, passing to the west of the Faeroe Islands at the halfway point. After about twenty days at sea, they reached Iceland.

As they approached the Icelandic shoreline, Christoforo would have noticed the wooden racks on the beaches where codfish hung on poles drying in the cold air. The racks were called stocks, hence the word "stockfish." He was familiar with eating dried salt cod, but this would have been the first time he had seen how it was made. Dried cod was Iceland's major export.

> "...fish like those which in Iceland are dried in the open and
> sold in England and other countries, and these fish are called
> in English 'stockfish.'"

JOHAN DAY

Since Bristol merchants had established trade connections in the Icelandic port of Snaefellsnes, this is most likely where their cargo was

20

unloaded and traded for local goods or cash. The new cargo was then carefully marked with the trademark of its owner.

Legally, the ship's purser was not supposed to purchase fish in Iceland. In 1475, in an international dispute with England, Denmark's King Christian decreed that England could no longer buy salt cod or stockfish from Icelandic merchants. Some trade was still taking place, but it was severely curtailed. It caused the merchants to begin searching for other fishing grounds.

The ship sailed 400 miles further north to the Arctic Circle, staying within sight of the Greenland coast, and then east to the abundant fishing grounds near San Jan Maya Island. Here, the crew engaged in fishing activities. When the cargo holds were filled with cod, they headed back south along the same route, the fish being gutted and packed in salt while the ship was underway.

Ships routinely making this voyage followed the same route every time, keeping within the sight of land as much as possible. They would risk leaving the sight of land for manageable periods of a few hours at a time, watching for the next landmark. It was a particularly cold watch up in the crow's nest, looking for signs of land. Communities would often maintain fires on coastal promontories to assist the seafarers, particularly when they relied on them for supplies.

When the ship reached the southern end of the west coast of Scotland, it veered west along the north coast of Ireland, stopping at the port of Galway, a small city on Ireland's west coast with a population of 4,000. Galway was a major destination for Bristol ships, both large and small.

In Galway, much of the cargo was transferred to Portuguese ships. These vessels traveled there to buy fish, which satisfied the insatiable demand for bacalao, or codfish, in Portugal.

The headquarters for sailors visiting Galway was the Church of St. Nicholas of Myra, and the Bristol and Portuguese crews, including Christoforo, visited the church to pray for a safe voyage before they resumed their journeys.

The voyage they had just taken to the north Mare Oceanum was a dangerous one, especially in winter. It would have been cold, wet, and stormy, and Christoforo must have wondered to himself why he had left the warm southern climes of his youth. He had firsthand experience with maps and sea charts, but he had stepped out of the confines of the Mediterranean by reaching this far north. Ice buildup in the rigging made these sailing ships unstable, and a trip of this magnitude involved a high degree of sailing ability and navigational skill.

Christoforo returned to Lisbon on one of the Portuguese ships, his original vessel returning to Bristol with one less crew member. In Lisbon, he was introduced to the chart making community, where his talents were soon put to good use.

The young Genoan sailor who had joined the crew in Bristol on the voyage to Iceland in February 1477 was little known at the time. But fifteen years later, he would achieve a position of renown and notoriety beyond imagination. His journals would be preserved for posterity.

4

Bristol Ships in Lisbon and Huelva

"Ships belonging to Bristol in the year of Christ, 1480. The Mary Grace 300 tons, le...of 360 tons, the George 200 tons, Kateryn 180 tons, Mary Byrd 100 tons, Christofer 90 tons, Mary Shernman 54 tons, Leonard 50 tons, the Mary of Bristow,...le George,...the John 511 tons, a ship that is just fitted for sea, John Godeman hath of ships,...Thomas Straunge about 12"

WILLIAM WORCESTRE, BRISTOL, 1480

L isbon, the capital of Portugal, had been home to Christopher Columbus for only a few months prior to his trip to England and Iceland. He had arrived there from his native Genoa in the summer of 1476. The son of a middle-class merchant family, Columbus had been at sea since leaving home at the age of fourteen. He had sailed the Mediterranean on small ships from Spain to Turkey, where he honed

23

his navigational skills. Furthermore, he had a keen interest in charts and cosmography and sought out the leading cosmographers to learn from them.

During political upheavals in Genoa in 1475, he joined a Genoan merchant ship, one of a convoy of five ships that was sailing from Genoa to England. A French pirate ship off the coast of Portugal attacked the convoy. The ship that Columbus was on caught fire, and the crew had to abandon ship. Columbus managed to cling to floating debris and swim ashore. He then made his way to Lisbon.

He and the other survivors were soon collected in Lisbon by another Genoan vessel sent to replace the lost ship. This ship sailed to London or Southampton and then to Bristol where it spent the winter. This vessel inadvertently introduced Columbus to a city that would play a central role in his career.

Columbus was an ambitious and intelligent young man. He had sailed the full extent of the Mediterranean Sea, and in his early twenties, he wanted to sail to the known limits of travel to increase his knowledge of the world and international trade. His curiosity encouraged him to travel as far north as possible.

Sailing to Iceland via Bristol afforded Columbus the most reliable and easiest way to sail to the Arctic reaches. The port of Bristol is situated about halfway between Portugal and Iceland, and was the only major English port with exposure to the Mare Oceanum.

Columbus's visit to Bristol in 1476 was a pivotal event in his career. Whether he visited Bristol by accident or by design is unknown, but the merchants he encountered there were the catalyst that would inspire him to make his discoveries fifteen years later. After he returned to Lisbon in 1477, Bristol trade and shipping took on a new significance. He now knew several Bristol ships and their connections with the Icelandic fish trade. He may have known some of the merchants that had dealings with them.

The *Trinity*, the largest ship in Bristol, sailed regularly to Lisbon, southern Spain, and North Africa, and the *Christopher*, a smaller vessel, provided a link between Lisbon, Bristol, and the community of Snaefellsnes in Iceland.

A short while after he returned to Lisbon from Iceland, Christopher was joined by his younger brother, Bartholomew, and they both worked as chart makers.

By 1478, the *Trinity* was in its fifteenth year of delivering Bristol's high-quality woolen cloth[1] to the Spanish and Portuguese markets through the ports of Huelva (Seville's port at the mouth of the River Tinto) and Lisbon. The ship would usually make the trip twice a year, trading the cloth mostly for wine, but also for a variety of food and general merchandise. Short stops would often be made at Kinsale in Ireland, Santa Maria (the port of Cadiz), Gibraltar, and occasionally Oran on the North African coast.

When the *Trinity* arrived in Lisbon, it would stay in port for three or four weeks. The cargo was unloaded and warehoused, and Lisbon's merchants would barter with the ship's purser for the goods. Many valuable relationships had been established over the years with local businesses.

Portugal's primary import from Bristol was woolen cloth, but leather shoes and gloves, glassware, and manufactured metal goods were also in high demand. Additionally, there was a market for the barrels of salted codfish and stockfish that Bristol merchants imported from Iceland.

Wine was Portugal's major export, up to 200 tuns a month. Olive oil, cork, wax, salt, sugar, fresh fruit, and other foodstuffs were also purchased in large quantities. Visits to the port of Lisbon yielded valuable cargoes for the Bristol traders and could be worth more than £1,000 per payload, almost as much as the value of the ship itself. Several English merchants lived in the Portuguese capital and acted as agents for their employers in Bristol. Columbus and his brother probably knew some of them.

The visit to Huelva, Seville's ocean port, was even more important to the Bristol merchants than the visit to Lisbon, and the *Trinity* would stay there for about a month. Seville was by Spanish law the major port of entry for most goods imported by sea from Northern Europe into Spain. As in Lisbon, woolen cloth was traded for local produce. The estate of the Duke of Medina Sidonia, the leading Spanish nobleman in Andalucia, was a regular supplier of large quantities of wine to the Bristol market.

During their stay in Huelva, the captain and the crew of the *Trinity* often called at the Monastery of Santa Maria at La Rabida. Accommodations were available there for sailors. One of the friars, Father Marchena, was an aspiring humanist and an expert in astronomy, astrology, and cosmography. He was particularly interested in learning from the English mariners about the voyages to Iceland and the Arctic, and the rumors of islands to the west of Iceland and Greenland. Before leaving port, the captain would settle the account with the head friar, Juan Perez, and make a donation to the monastery for the friars and the nuns in the convent to pray for the safety of the ship and the crew.

☩

Legends were told about large islands to the west, far out in Mare Oceanum, and Christoforo would have heard them. Far to the north, English ships traveled from Bristol to the Faeroes and Iceland in 1424. Icelandic fishermen and adventurers had traveled to Greenland and even further west to other islands. Christoforo may even have been aware of the Vinland map, which was drawn in Iceland around 1440. This map depicted what are now known as Greenland and part of the coast of the Canadian province of Labrador.

In addition to Iceland and Greenland, he would have heard of the Island of Seven Cities and the island of Brassyle. It was believed that the Island of Seven Cities was located approximately 1,000 miles west of Portugal. This island occurs on all Portuguese maps of this period dated later than 1400, indicating the confidence they had in its existence. Brassyle, 1,000 miles to the north, sometimes could be seen as a glow in the sky from the west coast of Ireland. Brassyle, also called Hi-Brassyle or Brasil, means in Gaelic the "Isle of the Blest." It was reputedly a tranquil place where the departed would spend time between the hardships of earth and heaven. Hi-Brassyle first appeared on the Angelino Dulcert

A CARAVEL, OR CARRACK

Chart produced in 1325. Originally thought to be about 100 miles west of Ireland, its suspected location gradually moved further offshore as English ships made periodic attempts to find it without success.

Several islands beyond Africa in the Mare Oceanum were discovered in the fifteenth century as mariners began to travel further from the sight of land. The Canary Islands had been known for 200 years, but Madeira was not discovered until 1419. The Azores were first sighted in 1439, but it was 1452 before all the islands in the group were accounted for. By the 1440s, Portuguese ships were visiting the Guinea coast of Africa, and the Cape Verde Islands were discovered. A Portuguese ship would first cross the equator in 1471 and round the southern tip of Africa in 1487.

✠

The substantial increase in the number and successes of the voyages of exploration in the 1400s was due to the introduction of the "caravel." This new design of ship made travel to the west coast of Africa and Iceland possible.

The caravel was a departure from earlier designs, with a high forecastle being added at the bow of the vessel and a similar aft cabin. Usually the sails and rigging consisted of three masts, one having a separate near-horizontal pole lashed to it with a triangular "lateen" sail attached. The other two masts were rigged with a yardarm and standard square sails.

The lateen sail enabled a ship to point about five degrees into the wind, something that had not been possible before. Sea captains could now take much longer voyages into uncertain waters, leaving the sight of land with the confidence that they would be able to return to their home ports by tacking and sailing into the wind.

The ships that predated the caravel were adequate for the Mediterranean Sea and European coastal waters between England, Germany, Holland, Northern France, and Ireland. They usually sailed in moderate conditions and seldom out of sight of land.

It was inevitable that these improvements in ship design would originate in Lisbon, the major port that is directly exposed to the Mare Oceanum. Small Portuguese ships were soon traveling up and down the coast to Spanish ports and exploring south along the African coast.

Shipshape and Bristol Fashion

Herring of Sligo and salmon of Bann,
Has made in Bristol many a rich man.

MEDIEVAL BRISTOL PROVERB

T he Bristol ships that captured Columbus's imagination would have been prominent on the Lisbon waterfront. They were more seaworthy than ships of other nations and integrated many of the latest advances in design. In the late 1300s, Bristol shipbuilders copied the Lisbon caravels. They began building larger vessels that carried more profitable cargoes. The Irish trade and later the Icelandic voyages became possible due to these dramatic changes in ship design. Bristol ships soon became noteworthy for their advanced design and construction.

The introduction of the caravel transformed Bristol from an English regional center to an international port. The rapid growth in exports of locally produced goods and the increased imports of Mediterranean foods and wine[1] created not only tremendous wealth and diversity in the city but also a new wealthy class, that of the "international merchant trader."

About ten Bristol ships regularly sailed to southern Spain and Portugal on voyages lasting from three months to as long as one year. The largest ship sailing from Bristol in the late 1470s was the *Trinity*, a caravel of 360-tun capacity and 100 feet in length.

The extreme tides in Bristol's harbor contributed to its preeminence in shipbuilding. Ships tied up at the harbor quays were effectively beached twice a day, often full of cargo, a serious event that put extraordinary strain on the hulls, necessitating stronger structures. The expression "shipshape and Bristol fashion" originated because of this challenge. If care was not taken when unloading a cargo, the ship could suddenly shift as the tide fell. This could be devastating; shifting cargoes could injure people inside and severely damage the ship. Ship owners favored one area of Bristol's harbor because the river bottom was muddy, and the hulls could rest easily in the mud.

A ship typically had a seaworthy life of between ten and fifteen years. Maintenance was continuous, and a ship's carpenter was a mandatory crewmember who would be employed to caulk leaks and replace wood structures, sails, and ropes during most voyages.

Due to the tides, dry docks were easy to build. Ships could be dry-docked on every tide. Consequently, ships were well maintained.

Strategically located on the west coast of England, Bristol is midway between major trading centers in Iceland and Spain and Portugal, and it offered safe anchorage for its fleet of merchant ships on a protected, easily defensible river. It had developed into a thriving commercial center serving the west of England, which it remains to this day.

Most of the shipping traffic in Bristol initially consisted of Welsh and English coastal traffic, but trade with Ireland had gradually grown in importance. By 1476, there were well-established trade connections with Kinsale, Cork, Limerick, and Galway. This regular trade with Ireland was the catalyst for Bristol's position in the history of exploration and

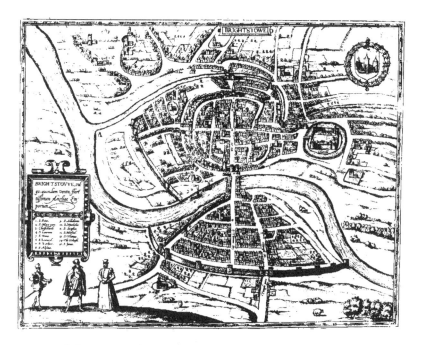

HOEFNAGLE'S 1581 MAP OF THE CITY OF BRISTOL

shipping. It provided the infrastructure, superior ships, and experienced sailors needed to sail north across the Mare Oceanum to Iceland.

It is not surprising, therefore, that Bristol ships, driven by the prospects of wealth for their owners, would be at the forefront of exploration and trade expansion.

✠

Bristol was England's second largest city and one of the world's major seaports. The port of Brygestowe ("bryge" means bridge and "stowe" is a place of assembly) is derived from the tenth century. Soon after the Norman invasion in the 1080s, William the Conqueror gave it a castle, and the city was granted a seal in 1350. Although decimated by the Black Death, by 1425, then known as Bristowe, it had grown and thrived due to its favored location.

Located 120 miles west of London, Bristol is surrounded by rich farmland, which supports dairy, crop, and sheep farming. Hundreds of small ships delivered wool, hides, and coal from South Wales, iron and oak trees from the Forest of Dean in Monmouthshire, and tin and fish from Devon and Cornwall. The River Severn is a major artery extending into the center of England, along which ships supplied grain and wool from Gloucester, Worcester, Hereford, and as far away as Coventry and Ludlow.

Extensive cottage industries developed, processing the local agricultural produce and the raw materials arriving by ship.

The city of Bristol in 1470 was small and compact. Most of the city was exactly as shown on Hoefnagle's map published in 1581, and the center of the city remains essentially the same today.

The city was built within a large bend in the river Avon, which meandered over a flat area of marshland before reaching the gorge. Protected by the river on three sides, a high city wall completed Bristol's defenses. The city was contained in an area between St. Augustine's Abbey and the Castle and covered little more than a square mile. Ten thousand people worked in the countless small workshops clustered in the narrow streets and alleyways around the older Castle area, on Redcliffe Street, and along the banks of the river.

The local production of large quantities of high-quality woolen cloth increased dramatically in the early 1400s, and Bristol came to dominate the cloth market in England for 100 years. Edmund Blanket owned one of the largest weaving factories. The manufacture of woolen cloth was the major industry, but a number of other small manufacturing industries had developed. Glassware was made in caves in the red cliffs by St. Mary Redcliffe Church. Hides were imported and made into shoes and gloves, and many metal shops manufactured tools, guns, church bells, and hardware. Both coal and lead were mined locally.

The docks constantly bustled with activity, most goods being moved by water even within England. There was constant activity all along the quays as the dock workers unloaded the cargoes and loaded other merchandise on board. Sixteen hundred sailors worked for Bristol shipping companies, and they and the dock workers patronized the numerous taverns on the crowded waterfront. Bristol's skyline was one of rooftops huddling together, smoking chimneys, tall sailing masts along the river, and the towering spires of her many churches. The church was powerful and supported education and academic life for those who could take advantage of it.

The commercial center, which lay between the Broad Quay and the Back Quay, reflected the wealth of her trade and was full of imposing buildings and business properties. This was the vibrant city that Columbus found when he arrived in Bristol in 1476.

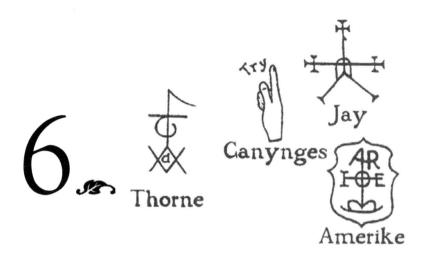

6. Thorne

Canynges

Jay

Amerike

The Fellowship of Merchants

"at capfinister go your cours north north est,
And ye gesse you ij parties ovir the see
and be bound into sebarne ye must north
and by est till ye come into Sowdying."

RUTTERS DIRECTING A SHIP FROM
CAPE FINISTERE, PORTUGAL,
TO THE SEVERN ESTUARY AND ON TO THE SHALLOW
WATER (SOUNDINGS) NEAR BRISTOL

Sturmy's House, a large former residence, was located on the quay by
Bristol Bridge near the Church of St. Nicholas. Robert Sturmy, a
wealthy merchant, gave his residence to the city to be used as the
Cloth Hall, the exchange where the woolen cloth makers traded their cloth
with the merchants for shipment overseas. A portion of the building was
used as the meeting place of an organization known as the "Fellowship of
Merchants."

35

The Fellowship of Merchants was founded in 1467 by Bristol's wealthiest and most politically powerful merchant traders. Their goal was to organize trade and the local shipping industry to their own advantage. They pioneered the formation of cartels to finance the construction and ownership of ships, consolidate cargoes, and generally look after their collective interests. The organization paid to have fires maintained on many headlands and islands, and was probably already administering charities to look after the dependants of seamen lost at sea. Through their membership, they had important connections with the English government and the King. This Merchants' Guild, as the Fellowship was also known, was run in parallel by the City of Bristol government. The Mayor and City counselors made all appointments.

Many ships were still owned by individual merchants in the traditional way. The Fellowship, however, made it possible for several merchants and investors to band together to build and operate a merchant vessel. Much larger ships could be built when the financial risk was spread over several partners.

A new business class evolved—that of the ship owner. These ship owners solely operated ships and leased vessels or cargo space to merchants for specific voyages or time periods. They provided a ship that came complete with a captain and a crew, and the return to the ship's owner was usually tied to the profits made on the voyage. The profits were such that a caravel could be paid for in just four voyages.

Merchant trading companies leasing the ships had only the responsibility of assembling cargoes and trading them in foreign ports.

The *Trinity*, a magnificent 360-tun caravel, was built in 1463 and regularly sailed to Spain and Portugal. It had cost about £1,200 to build and was one of the finest ships in England. It was built by a cartel of eight Bristol merchants that included John Jay, John Shipward, William Bridde, Gilbert Smyth, and William Wodyngton.

Jay was the sole owner of several ships and had business interests in many others. In 1477, he placed five ships at the disposal of King Edward

VI, and he was allowed to import goods to the value of £190, duty free, in return.

The shares of ownership in a vessel were readily transferable, and on John Jay's death in 1468, his shares passed to his sons, John and Henry.

In one of its earlier voyages, the *Trinity* delivered 700 cloths to Spain and Portugal, the equivalent of three miles of cloth three feet wide. This was an enormous cargo for a ship at that time.[1] As many as 100 different merchants consigned cargo on each voyage, the accounts for each merchant being kept separately by the ship's purser.[2]

Columbus may have known the *Trinity*. It regularly traveled from Bristol, stopping in Spain, Portugal, North Africa, and Ireland, and would have been a familiar sight in Lisbon Harbor in the late 1470s. It had a crew of 30 men, and their pay was typically 12d a week. The operating costs for the ship were about £8 a week.

The resident purser in the late 1470s was a man named John Balsall, and his trading records for one particular voyage were discovered in London just forty years ago.[3]

Many names listed on the bills of lading are familiar to scholars of Bristol's history. John Jay III, William Bird, Robert Strange, John Esterfeld, William Wodyngton, and Richard Amerike were just a few of the merchants who owned the cargoes of woolen cloth on board. They were all members of the Fellowship of Merchants, as were the owners, John Jay II and Thomas Croft. According to the Balsall records, a few women merchants also shipped cloth on the ship.

Some of the names in these trading records would soon be repeated in an English crown charter, many appear on the list of Bristol's Mayors and Sheriffs, and one name in particular would become prominent twenty years later. Messrs. Spencer, Pynke, Esterfield, Strange, Thorne, and Amerike had served terms as Mayor or Sheriff of Bristol, or they would in the future.

By Balsall's accounts, the *Trinity* returned to Bristol six months later, laden with 76 tuns of wine, 182 tuns of olive oil, 53 cwt of sugar, 59 cwt of wax, and other produce. Richard Amerike was the importer of the sugar and the oil.

Thomas Croft was a business associate of John Jay II. He also held a one-eighth share in the *Trinity*, as well as shares in other vessels. In 1477, he was working as the King's Customs Officer in Bristol. This was an honorary appointment, and it was his responsibility to collect all customs duties for the Crown. Croft was from an influential family and had been a friend of King Edward since his childhood at Croft Castle, near Leominster in Herefordshire. The King was raised at nearby Ludlow Castle, and the families were close both socially and politically.

Customs duties were used partly to finance the Royal Navy. Piracy was a major problem in the early 1400s, so merchant ships would travel in convoys, accompanied by one or two Navy vessels for protection.

Another Bristol merchant starting to achieve prominence, and a generation younger than John Jay II, was a man named Richard Amerike. He was a wealthy landowner and merchant trader who, by 1478, was building a successful business trading with merchants in Spain and Portugal. He had built up a trade in exporting high-quality woolen cloth. Only forty years old, Amerike was a contemporary of John Jay III and was likely in a similar position: that of running his family's business. Given the connection between these influential families, it is possible that Amerike or his family also owned a share of the *Trinity*.

Amerike lived with his wife, Lucy, and their two teenage daughters in the Clifton Manor, one third of which he purchased in 1470 from Sir John Chideock. Amerike's land consisted of about 100 acres of the Clifton wood and the adjacent farmland centered around St. Andrew's Church and Clifton Green. It included a manor house, many dwellings, and several small tenant farms. The property is situated at the top of a steep hill overlooking the river and the lands of St. Augustine's Abbey. Nearby are the cliffs and the steep gorge through which the trading ships sailed to reach the city's port.

Amerike's family was of Welsh aristocratic descent.[4] They had been wealthy landowners in the Welsh border country near Ross-on-Wye in

Monmouthshire for many centuries. The anglicizing of his surname from Ap Meric to Amerike was still not fully established.[5] This might indicate that the family's move to England was very recent or, more likely, that the family was resident in Bristol and still had an estate in Wales. Perhaps Amerike's father or grandfather moved to Bristol in the early 1400s to trade wool, which was regularly shipped from Wales to be made into cloth in the Bristol workshops.

Richard Amerike was also a contemporary of Thomas Croft, who was raised on his family's estate in Herefordshire, just a short distance north of Ross-on-Wye. They probably knew each other since childhood.

Richard Amerike was one of the wealthiest of about 250 individuals living in Bristol who were trading internationally in the late 1470s. These were not all full-time merchants, but there was a group of about thirty-five wealthy men whose main business was exporting, trading, importing, and distributing foreign goods, investing at least £100 per assignment.

The Spanish and Portuguese trade in which Amerike was becoming a dominant figure had only existed for a little over a decade but was growing rapidly. Prior to 1455, few English ships had ventured past the French city of Bordeaux in the area known as Gascony. Bordeaux, like much of France, had been an English possession and was the principal source of wine. In 1453, the French reestablished control of most of France, and trade with Bordeaux abruptly ceased.

This event exacerbated the Wars of the Roses, which pitted the House of York against the House of Lancaster for control of the English Crown. The Yorkists were ruling at this time when Amerike's wine trade was interrupted, and he was forced to venture into the Bay of Biscay. An uneasy and fragile peace with France was arranged. This apparent peace did little to quell the violent English civil conflict that resulted and was prone to flare up periodically for more than thirty years.

Richard Amerike's family had been prominent merchants in the Bordeaux trade. An early record shows John ap Meric and Richard ap Meric importing wine on a ship called the *George* in 1436. Richard ap Meric may have been Richard Amerike's father.

With the collapse of the wine trade from France, new sources were urgently needed. Improvements in the design of ships built in the Bristol shipyards enabled the merchant seamen to give Gascony a wide berth and sail further south and across the notoriously rough seas of the Bay of Biscay to Spain and Portugal.

Amerike would come to dominate the expanding trade with Spain and Portugal, which by 1479 had grown to over half the value of all foreign trade in Bristol. He shipped out primarily woolen cloth and imported wine and many other commodities.

With as many as several dozen individual merchants each owning cargo on any particular voyage, a method was devised to enable the purser to keep track of each merchant's property. Merchants had trademarks that were used to identify their cargoes. This mark was chalked onto the barrels or cargo. Each was distinctive in appearance and was recognized by the Fellowship of Merchants. This same mark was engraved on a ring and used to seal documents with a wax seal.

The ship's purser was authorized to trade the cargo in foreign ports. His sole judgment was relied upon to barter a merchant's cargo for local goods if the merchant did not have a local representative, which was usually the case. The trademarks were an effective way of identifying the cargo when ships were being loaded and unloaded and when the purser was trading.

Most merchants confined their trading to specific geographic areas. There was a growing and increasingly more affluent group that traded more or less exclusively with Spain and Portugal. They bought large quantities of woolen cloth for export, confident that they had a market for it.

Another distinct group concentrated on the northern countries of Iceland, the Baltic region, and Ireland. The commodities traded were quite different. Many were fish merchants, and most cargoes were of considerably less value than the cargoes originating in Spain and Portugal.

✛

One of the most important functions of the Fellowship was to maintain the "Map Room." Before the advent of the printing press, there were no charts available except ones that were hand drawn. The chart maker in Bristol drew these maps and constantly kept them current. They were maintained and kept in a secure place.

Ship captains did not have charts but were supplied with "rutters" describing their assigned voyages. These documents were taken on board in the captain's leather satchel and stored in a safe, dry place when not in use. The captain did not have the luxury of a dry cabin to work from and was usually on deck, directly exposed to the weather.

When the captain returned to Bristol, any new information he could add was given to the chart maker in the Map Room. If a new island or destination was discovered, rutters would be written or amended and made available for ships sailing there in the future.

Seamen have always had local names for the marks and headlands they looked for as the ship approached land. Names of places originated in many uncharted and obscure ways, for example, after a particular geographical feature, a family name, a business, or a profession. New names quickly take root through verbal use, and these verbal christenings have justifiable staying power.

Discoverers, by contrast, often named landmarks after dignitaries or members of the ship's crew. The captain of the ship entered these names into his charts and there they remained. Captain George Vancouver named most of the landmarks in Puget Sound and around Vancouver Island after English royalty, naval dignitaries, and his ship's officers. Puget Sound was named after Peter Puget, one of his officers; Mt. Rainier was named after the admiral; Mt. Baker was named after a junior officer; and, of course, Vancouver Island was named after himself.

7.

A Fish Story

Salt Cod

"The sea is swarming with fish, which can be taken
not only with the net, but in baskets let down with
a stone...his companions say that they could bring so
many fish that this kingdom would have no further
need of Iceland, from which place there comes a great
quantity of stock-fish."

RAIMONDO DI SONCINO, MILANESE AMBASSADOR,
LONDON, 1496

There was a staggering demand for fish throughout Europe in the 1400s. For those who lived on the coast, fresh fish was usually available. However, for people who lived inland, especially in hot climates, the supply of fresh fish was unreliable and expensive or even nonexistent. Members of the Catholic Church created a particular demand for fish on Fridays and all holy days, not only in England, but especially in Spain and Portugal.[1]

Norwegian fishermen used an effective process for drying and preserving fish. They exported the process to the Icelandic people and the settlers in Greenland. They originally produced "stockfish" and "salt cod" for their own consumption, harvesting and curing it for use when bad weather made fishing impossible. This method was later used to meet the demand for export. The process works best with cod because cod has a longer shelf life than other salted fish, and it rehydrates and tastes better when reconstituted. For the poor, this dried fish was cheap, high-quality nutrition with a long shelf life.

The fishermen started the process for preparing stockfish immediately after the catch. The fish were gutted and heads were removed. They were filleted leaving the two halves joined at the tail so that they could be hung over poles and placed on stocks that the fishermen built on the beaches. The fish were dried in the cold, dry air for as long as three months, and they would shrink to a fraction of their original thickness and lose eighty percent of their weight. When completely dry, the fish were tightly packed into barrels between layers of salt.

Salt cod was made by compressing fresh fillets of cod between layers of salt. Both types of fish kept for long periods of time, and large quantities of fish could be transported to hot climates without spoiling.

When stockfish was brought to market, it was soaked in a bath of water for many hours. The water was refreshed several times to eliminate most of the salt. As the fish absorbed the water, it swelled to several times its original thickness. The fish was then poached and cooked in sauce. It was a tasty and popular food.

Stockfish is produced in Norway and Iceland today, and there is still a thriving export market. The Spanish have a dish called baccala. In Portuguese, it is bacalao, and it is prepared in the same way.[2]

✠

In the first half of the fifteenth century, a few Bristol ships traded with Icelandic and Greenland settlements, and the Danish-Norwegian authorities reluctantly allowed them to continue trading when they increased their regulation of the stockfish market. Shipping was supposed to go through the Norwegian city of Bergen, where prices and taxes were controlled.

Bristol merchants sent a large number of vessels directly to Iceland and Greenland every year; enforcement of the regulations was lax. The English ships paid twice as much as the Norwegian buyers for the fish, so they were welcome in the small coastal settlements. Most of the larger 200-tun ships sailed north through the Irish Sea between Wales and Ireland, but some would sail into the Mare Oceanum putting into the Irish port of Galway. There were also many smaller coastal traders of about 50-tun capacity, called doggers, that plied this route with lower valued cargoes, such as wood, that they traded for fish.

The *Anthony* made a typical voyage to Iceland in 1471:

> "*The ship called the Antony of Bristol in which John Deanfitz is master came from Island on this day (September 10) and has in it for John Forster, denizen, xxxv last lying, value lxx pounds, subsidy lxx shillings for the same, xv last stockfysshe, value lxxv pounds, subsidy lxvv shillings*
> *for John Gregorie, denizen, j last salt fish, value x pounds, subsidy x shillings.*"

A last was a unit of measurement, 640 gallons or the volume of a box 4' x 4' x 4'. The value of this cargo was as follows: thirty-five lasts (probably of fish soaked in lye, or lutefisk) £70, fifteen lasts of stockfish £75, and one last of salt cod £10. On each, a subsidy, or duty, was payable of one shilling in the pound, or five percent. A typical value for a cargo in this market was £155. By way of contrast, the value of a cargo on one of the larger ships from Lisbon or Spain was at least £1,000.

The major commodity needed to make stockfish was salt. The Portuguese had this in abundance, and they also had a huge market for the stockfish. The Bristol merchants had the seagoing ships, so they established a regular trade route in the mid 1400s. Bristol ships returned from Portugal in the spring loaded with salt. Salt was then shipped to

MAP OF THE NORTH ATLANTIC

Iceland to be traded for stockfish. The ships returned directly to Bristol or sometimes stopped in Galway, where the fish was unloaded and shipped to Portugal and Spain.

The ships trading from Bristol in the Icelandic and Baltic markets usually did not venture into the Spanish or Portuguese market. However, there was one exception. William de la Founte was involved in trade in Iceland as well as in Spain and Portugal.[3]

A small Icelandic community lived in Bristol. They mostly worked for the wealthy merchants as household servants.[4] One Icelandic national living in Bristol was a man named Willelmus Island. He became a merchant and later specialized in the Portuguese market. He eventually became an English subject.

The Icelandic people were not wealthy, but they have a rich history and folklore of early exploration in the Arctic region. The major fishing grounds are to the south of Iceland, but Icelandic fishermen are known to have ranged as far west as Greenland and Labrador in their search for fish and timber. They established several new communities in the process. In the tenth century, Leif Eriksson sailed from Iceland and established a settlement in Labrador that existed for over 300 years. The winters were inhospitable, making it difficult to establish a permanent settlement, and it eventually fell into disuse. However, the fishing was better than anyone ever imagined.

The cold Labrador current flows south from the Arctic carrying minute shrimp, called krill, that fish feed on. The current meets the warm Gulf Stream flowing northeast up the Eastern Seaboard off the coast of Newfoundland on a shallow area we now call the Grand Banks. This cauldron of cold Arctic water and plankton-rich warm water provides inexhaustible feed for fish. The St. Lawrence River also joins the ocean in this same area. These physical features in combination produce the best fishing area in the world. In July, smelt proliferate and their presence attracts codfish and whales. The cod can grow up to 5 or 6 feet long and weigh 200 pounds.

✝

Shipping in the entire Baltic region was controlled by an international organization called the Hanseatic League. The League had been formed in Germany in the thirteenth century to regulate trade and control the mouths of the major rivers in Northern Europe. It was initially seen as a positive force, but it gradually came to be dominated by Scandinavian and Northern European interests. The League regulated the fishing industry and gave licenses for the exclusive rights to sell fish in specific markets. In 1450, King Christian of Denmark, the head of the Hanseatic League, granted a Bristol company the exclusive right to import fish from Iceland to England. Then, in 1475, the Hanseatic League, at the instigation of the Danish-Norwegian crown, attempted to monopolize the dried cod trade and eliminate the English from the market. They enacted a doctrine to prevent English merchants from buying salt cod from Iceland.[5]

The dried fish market in Bristol collapsed overnight. A few English ships attempted to circumvent the ban, and there were several hostile incidents. Danish authorities seized some of the English ships. The loss of this lucrative trade had a devastating effect on the financial wherewithal of the Canynges Company.[6] It also created a serious problem for the Bristol merchants who specialized in trading in the southern markets. Their customers and agents in Lisbon and Seville relied on them to supply the considerable demand for salt cod.

The Bristol merchants would have to find new sources of fish.

✠

Interest began to center on the mysterious island of Brassyle that supposedly existed west of Ireland. Bristol ships traveling to Galway had made forays into the Mare Oceanum but had not been able to locate Brassyle. The distance from the Irish coast was thought to be at least 400 miles.

The Icelandic people living in Bristol related stories about the known lands to the west beyond Iceland and Greenland where exceptionally good fishing grounds were reported to exist. Willelmus Islond, the Icelandic merchant, was a member of the Fellowship of Merchants, and he would have been familiar with the stories. The merchants wondered if it wasn't Brassyle that the Icelanders were describing.

William de la Founte, William Spencer, and Richard Amerike each employed Icelandic servants who would have been familiar with Icelandic folklore and legends pertaining to these lands.

De la Founte was a half owner of the *Christopher,* which often sailed from Bristol to Lisbon and Bristol to Iceland. One of his skippers, an accomplished seaman by the name of Thomas Sutton, traveled from Gibraltar to Iceland and back, always passing through Bristol.

On a typical voyage, he would bring fruit, oil, and salt from Lisbon to Bristol, where only the fruit and oil would be unloaded. The salt and additional cargo would then be sent to Iceland. On the return haul, some of the stockfish would stay on board in Bristol to be sent south. The ban on fish sales in 1475 severely disrupted de la Founte's business.

Bristol merchants paid special deference to rumors that the Basques[7] in Northern Spain had discovered the island of Brassyle. Their commercial sphere of influence was in the area centered on the port of Bayonne, a city near the French-Spanish border on the Bay of Biscay. In the mid 1400s, Basque fishermen dramatically increased their fishing catches, and their neighbors to the north, the Bretons, began to notice their increased wealth. Other fishermen began to wonder why the Basques were so successful and where they were catching fish. The Basques refused to reveal the location of their fishing grounds, but it was getting increasingly more difficult for them to protect their secret.

The Bretons, fishing out of Brittany, tried to follow the Basques but were unsuccessful. However, their stories of an island across the Mare Oceanum persisted, and seamen from Bristol heard the rumors in the taverns and foreign ports.

Denmark's decree in 1475 encouraged the Bristol merchants to make a concerted effort to find Brassyle and even the Island of Seven Cities.

Croft

Straunge

Spencer

8

The Search for Brassyle

"Licence for Thomas Croft and for William Spencer, Robert Straunge and William de la Fount, merchants of Bristol, to trade for three years to any parts with any except staple goods, despite any statute to the contrary, with two or three ships, each of 60 tuns or less."

ISSUED BY EDWARD IV, KING OF ENGLAND
PUBLIC RECORDS OFFICE, LONDON,
TREATY ROLL 164 M. 10, 18 JUNE 1480

The merchants who traded with Spain and Portugal were the ones most concerned with the loss of the salt cod and stockfish trade. The resale of Icelandic cod to the rapidly growing Spanish and Portuguese market was financially more rewarding than to the English market, which was mostly met by imported fish from Ireland.

The Bristol merchants had long been familiar with the Greenland coast, and it is likely that they had reconnoitered a route from Iceland west and south around Greenland to Labrador or Newfoundland. The merchants had sufficient knowledge of the distant waters to approach the Crown for an official exploration permit.

In 1479, a delegation of the four most senior and influential merchants in the Fellowship applied to King Edward IV for a trading license to explore and trade in new territories.

To this effect, the King granted a trading license effective on June 18, 1480, to Thomas Croft, Robert Strange, William Spencer, and William de la Founte. It restricted the exploration to three ships, each with a maximum cargo capacity of sixty tuns. This meant that they could take three caravels, each about seventy feet long. They were authorized "to trade for three years to any parts with any except staple goods," and it stipulated that the Crown was to receive one-fifth of the profits after all costs had been met. The document would expire on June 17, 1483.

This license gave the merchants the backing of the English Crown to explore and trade in new areas and receive the protection of the English government and the Royal Navy. For this they would pay the Crown one-fifth of all profits. Foreign governments were made aware of the expedition, and ship owners were authorized to explore within the terms of the license.

The government official and the three merchants named on the document were prominent in Bristol civic affairs, and each owned ships or had businesses trading with Spain and Portugal.

Robert Strange, aged forty-two and the youngest of the four, was from a family that owned at least twelve large ships and was an established wine merchant. He was Mayor of Bristol in 1475 and would serve another two terms in 1482 and 1489. He also served two terms as its Member of Parliament.

William Spencer, aged fifty-seven, was the Mayor. He later served three terms in this capacity and a further term as Bristol's Member of Parliament. Spencer was the merchant who, with John Pynke, an ex-Mayor, consigned goods from Lisbon to Bristol on the *Christopher*.

William de la Founte was a wealthy merchant who owned a large 201-tun ship and was also half-owner of the *Christopher*.

Thomas Croft was the Customs Officer. He owned a one-eighth share in the *Trinity* along with several other partners, including John Jay. Croft would be responsible for collecting any money due to the Crown.

Two other younger merchants, John Jay III and Richard Amerike were most likely junior members of the cartel.

Meanwhile, the regular trade continued unabated, and the *Trinity* set sail from Bristol to Lisbon in October 1479 with a full load of woolen cloth.

During the winter, several other ships were loaded with similar cargoes bound for the Spanish and Portuguese markets. The outbound ships included the *George* and probably the *Christopher*, which sometimes traveled together for mutual protection.

The *Trinity* arrived back in Bristol on March 4, 1480, fully laden with cargo including sugar and olive oil belonging to Richard Amerike. Thomas Croft, in his capacity as the King's Customs Officer, could legally own the vessel, but he was prohibited from trading in goods.

During March and April, the ship was unloaded and then restocked with another cargo of cloth. It left Bristol in May 1480 for a similar trip to Lisbon. It eventually returned to Bristol about four months later. Mystery and intrigue still surround the events of this particular voyage.

One of the ship captains employed by Jay was a man named Lloyd. If indeed he was the skipper on this particular voyage, then the return trip to Bristol was not straightforward. The *Trinity* left Lisbon on July 15, 1480, and seemingly took a clandestine detour to look for a mysterious island in the Atlantic.

William Worcestre later wrote:

> [Bristol men] *"in two ships of 80 tuns, of Jay, junr. A merchant, who began their voyage 15 July 1480, at the port of Bristol at Kynroad, for the island of Brasyle, taking their course from the west part of Ireland, plowing the seas through, and Thlyde (Lloyd) is master of the ship, the most skilful mariner in all England…News came to Bristol Monday 18th*

September, that the ships sailed over the seas for nine months,
and found not the island, but through tempests at sea returned
to port in Ireland, for laying up their ships and mariners."

Worcestre was a prolific writer and documented much of the history of Bristol during his day. He was the brother of Joan Jay, the wife of John Jay, the Elder. Jay, Junior, was his nephew. This voyage from Lisbon commenced on July 15, 1480, just four weeks after the charter became effective.

Worcestre's writings describe two 80-tun ships that belonged to John Jay sailing from Bristol with Lloyd, "the most skillful mariner in England," as Master. They were trying to find the island of Brassyle, which was supposed to be 400 miles west of Ireland. After searching the Mare Oceanum for nine weeks, the ships did not return to Bristol but put into a port in Ireland, from where news reached Bristol via another ship. Subsequent events even raise the possibility that one of these ships may have been called the *Trinity*.[1]

Worcestre was quite old when he wrote this and was living in Norfolk, a county about 200 miles east of Bristol and northeast of London. He was writing his family's journals, probably using letters and stories that were told by members of his family. He obviously erred when he described the voyage as being nine months instead of nine weeks, and he may have confused the date the ship left Lisbon with the date it left Bristol. If the Jays were using the large 360-tun *Trinity* instead of a smaller 60-tun vessel allowed by the trading license, they would not want that fact too widely known.

In October 1480, the *Trinity* was back in Bristol and set sail in that month under the command of Master Rychard Parker for another uneventful voyage to southern Spain with a cargo of 400 bolts of woolen cloth. The young merchant Richard Amerike owned a portion of this shipment. It followed its regular trading route from Bristol to the Welsh port of Pembroke. From there, it sailed to Kinsale in Ireland and then with a full cargo to Huelva in Spain. It stayed in Huelva for a month and while there, John Balsall, the purser, traded much of the cargo for cash and goods. He recorded in his accounts unloading and warehousing forty whole cloths in his capacity as agent for the following merchants:

"*In primus y recevid by the grace of God out of the Trynety by ffore wreten at Welva ffor the name of Master Wylliam Spencer marchaunt of Brystowe ....conteyneng vii holl cloths......- 7 holl cloths*
Item......Master Straynge...... 5 holl cloths
Item......Master Esterfyld....... 7 holl cloths
Item......Master Edmond Wescott..... 5 holl cloths
Item......Master William Bryd & John Jay.... 11 holl cloths
Item- more y recevid of the seyd schep at Welva yn the name of Rychard Amyreke
marchaunt v holl cloths under the mark made in the margent - v (5) holl cloths"

JOHN BALSALL PURSER OF THE "TRYNETY OF BRYSTOWE" - HUELVA, CASTILLE, 27 NOVEMBER 1480. AN ACCOUNTING OF THE SALES ON DECEMBER 18, 1480.

This group of six merchants includes the same William Spencer and Robert Strange who had recently been granted the "license to trade - dated 17th June 1480," and also two younger merchants named John Jay and Richard Amerike. Amerike is listed last and is the only one of the group that Balsall does not defer to as "Master."

The ship then sailed further south, stopping in Gibraltar, and into the Mediterranean Sea to Oran in North Africa. On the return trip, in April 1481, the *Trinity* again stopped in Huelva, trading more cargo for wine and oranges before returning to Bristol. Cargo was traded at every port to maximize profits.

✝

No official records exist stating how many exploratory voyages the Bristol merchants made that summer in their search for new fishing grounds or of their success. They kept these voyages secret, as they had nothing to gain if word got out that they were successful. Their only concern was to find and import fish. Had it not been for Worcestre's personal writings, we would not know of any voyages made in the first year under the trading license.

There were likely other forays in that summer in 1480, and every indication is that one of the ships found Brassyle along with fishing camps that had been set up there by the Basques. If this is the case, the Bristol ship was there only to trade. The terms of their license did not give them any jurisdiction to claim the land for England.

In October 1480, there was at least one other voyage. This time the ship had a very unusual cargo.

One of the ships called *Trinity* set sail from Bristol. The destination was recorded as Ireland. Richard Amerike consigned on board a cargo of two tuns of salt and six tuns of corrupt wine. This is not the usual cargo that would be sent to Ireland, so it is unlikely that Ireland was the intended destination.

Salt would have to be imported to Bristol, most of it from Portugal. Consequently, it was expensive. This was a significant investment for Amerike and one that would have been planned in advance to include appropriate arrangements on the receiving end.

The following summer, in July 1481, the *Trinity* and another ship called the *George* were both in Bristol, and they were again loaded with salt and other supplies. Thomas Croft and John Jay owned shares in these ships.

The destination for these ships was also recorded as Ireland, but this proved not to be entirely true. The journey, which should have taken three weeks, took nearly three months.

These individual voyages were supposed to show a profit, and by this time the merchants were almost certainly bringing fish back to Bristol. There were, however, no records of any fish imports in Croft's customs accounts.

According to the terms of the license, the merchants should pay twenty percent duty on imported goods resulting from trade, after all costs were subtracted. Since their costs were high, they may have calculated that there was no reason to pay customs duty if these voyages were not showing a profit.

On the other hand, if they had established a fishing settlement and were catching the fish themselves, the fish would not have been acquired by trade and would therefore not be taxable. This possibility and the clandestine movement of large quantities of salt and corrupt wine strongly support the idea that Amerike, Jay, and Croft had indeed established their own fishing settlement. These events didn't go unnoticed in London.

Immediately after the *Trinity* and *George* returned to Bristol, suspicions were raised and questions were asked in London about the real purpose of the voyage. The government evidently believed that Croft and the Bristol merchants were trading and illegally importing goods, thus evading the twenty percent payment to the Crown. A Royal Commission was sent from London in September to examine their activities. In October 1481, Thomas Croft, the Customs Agent, was arrested and charged with engaging in foreign trade:

> *"Thomas Croft of Bristol...Customer of the said lord the king in the port of his town of Bristol aforesaid on the sixth day of July in the aforesaid year (1481)...was owner of an eighth part of a certain ship or balinger called the* TRINITY, *and an eighth part of a certain ship or balinger called the* GEORGE, *and in each of the said ships or balingers...laded, shipped and placed forty bushels of salt...with the intention of trading."*

Croft pleaded that the *Trinity* was exploring and not trading. He said in his defense that he was only trying to find Brassyle and that the salt was on board for the ship's "repair, equipment and maintenance." He did not dispute that Ireland was not the intended destination or deny that the ships were exploring; he only denied that they were trading. After lengthy court proceedings, including jury deliberations, he was acquitted.

It is far from certain that the real truth emerged at Croft's trial.[2]

Amerike, Jay, and Croft must have had a safe harbor in mind when their ships left Bristol. The first Bristol ship to discover Brassyle was indeed looking for trading possibilities as authorized by the license.[3] When they found Brassyle, they almost certainly located a safe harbor where they could go ashore and unload their cargo.[4] They may have traded with fishermen who were already there or, more likely, decided to set up their own fishing settlement. They probably built some wooden structures to protect the salt and other supplies from the elements.[5] This would be their base in Brassyle.

When the ship returned to Bristol, the captain reported to the Map Room at Sturmy's House. A map would have been drawn showing the location and the details of any safe harbors for future use.

Other ships were soon prepared for the voyage to Brassyle, and the masters and crews of these vessels would have been briefed in the Map Room and given their rutters and the bills of lading of the cargo they would be delivering. The rutters would accurately describe the location of the safe harbor with a detailed description of the route the ship should follow.[6] These men were merchant seamen delivering cargoes for their employers. They were not voyagers or discoverers.[7]

9.

The Winter of Discontent

"For the last seven years the people of Bristol have equipped two, three, or four caravels to go in search of the Island of Brazil and the Seven Cities."

PEDROS DE AYALA, THE SPANISH ENVOY IN LONDON, REPORTING TO THE KING AND QUEEN OF SPAIN, 1497

The years between 1483 and 1486 were tumultuous in England. The country had been in a state of civil war since 1453 when two branches of the English royal family, the House of Lancaster (represented by a red rose) and the House of York (represented by a white rose) fought each other for the English crown.

King Henry VI of the House of Lancaster ascended to the English throne as a baby in 1422. He was a weak king and subject to periods of insanity throughout his life. His mother was French, and he became the King of France in 1430. Joan of Arc led an uprising in France, culminating in 1453 when the French drove the English out of France except for the Channel port of Calais. This event caused the collapse of the trade between Bordeaux and Bristol.

For two years after this event, King Henry was given to bouts of insanity, and his distant cousin, Richard, Duke of York, reigned as Protector with the King's infant son. In 1455, York was dismissed, and war broke out briefly between the Yorks and the Lancastrians.

Over the thirty year period ending in 1487, five different kings would reign through seven changes of monarch. Three of these kings would die violent deaths.

In 1460, war broke out again, and in 1461, Henry was defeated by Edward of the House of York at the battle of Mortimer's Cross. This battle was fought on land owned by the Croft family near Croft Castle, Leominster, Herefordshire. Henry was deposed, and Edward was crowned King Edward IV. In 1464, the King married Elizabeth Woodville in a controversial union in that she exerted undue influence over him and state affairs. Her sister was married to Sir John Poyntz in Bristol.

In 1469, Edward was defeated at Banbury, and he fled to France. In October 1470, the Earl of Warwick reinstated King Henry VI. Five months later, Edward IV returned, and after the battles of Barnet and Tewkesbury, he again claimed the throne. Soon afterwards, Henry was stabbed to death.

Edward IV revived the English claim to the French throne and invaded France in 1475. He extorted a nonaggression treaty with King Louis XI and exacted a large annual payment from the French.

He died suddenly in 1483 at the age of forty-one and was succeeded by his 13-year old son, Edward V. The child was met by his uncle, Richard of York, was taken to London, and was held in the Tower of London from where he disappeared. Sir James Tyrell murdered him on the orders of his uncle, who was then crowned King Richard III.

The Yorks were allied with the aristocracy in Brittany and sought to prevent the French King's territorial expansion into that area. The Breton aristocracy looked to England for assistance.

Fifty years earlier in the 1430s, a son of King Henry V, the Lancastrian Owen Tudor, had lead an unsuccessful uprising in Wales. His family lost their extensive land holdings in Wales as a consequence.

Owen Tudor's grandson, Henry, the Earl of Richmond, was raised until the age of fourteen in Pembroke Castle in South Wales and was then expelled to France with his family. He saw an opportunity to take the throne and regain his family's land and power. In 1485, he felt that he could gain enough support, particularly then, when King Richard was in power. Richard's only child and heir had died in 1484, and his wife died in March 1485. Henry succeeded in obtaining the support of the French King, and in turn he supported French territorial expansion into Brittany.

In the spring of 1485, the Earl of Richmond landed with an army at Milford Haven in South Wales. He marched across Wales to Shropshire, in the English midlands, and defeated King Richard III at the Battle of Bosworth Field. Richard was killed, and the victorious Richmond was subsequently crowned King Henry VII.

This was the end of the Plantagenet era and the beginning of the Tudor dynasty. King Henry VII claimed descent from the ancient British kings and King Arthur through his Welsh grandfather, and he adopted the Welsh dragon as one of the supporters of the royal arms. This was a good time to be a Welsh aristocrat.

This decisive date also marked the end of the medieval period in England and the start of the Renaissance, a 150-year period of unprecedented prosperity and sophistication.

✠

In Bristol, it was a favorable time to keep a low profile. The shrewd merchants were politically astute and survived well the changes in royal succession.

The Bristol merchants continued to secretly deploy a small number of ships to Brassyle for dried salt cod. Sixty to one-hundred tun fishing vessels were loaded with salt and supplies in Bristol and recorded in the

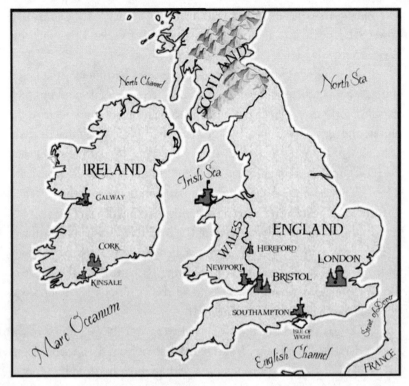

MAP OF GREAT BRITAIN AND IRELAND

port records as destined for Ireland, but they did not return to Bristol for several months. Fishermen are known to protect their fishing grounds when they find a lucrative area.

Navigation techniques for the open sea were evolving and improving, along with ships that were able to take increasingly longer voyages in unknown waters.[1] Charts were not available to English captains. The ships would travel north for 800 miles and then west, skirting the south coast of Iceland and reaching for the southern coast of Greenland. The final stretch of open water was a day and a half sailing (in favorable conditions) to the Labrador coast. The captain could measure latitude fairly accurately, and the ship would cross open sea, holding firm to the same line of latitude to landfall.

The discovery of Brassyle was not exactly the trading opportunity that the Bristol merchants had hoped for. There is no doubt that the fishing was good. The Bristol merchants had never seen such rich fishing grounds. However, there was virtually no market for the wool cloth, the hardware, and the food they hoped to trade with the fishermen selling the stockfish. The natives in the new land had nothing to trade, and the only supplies they could sell were salt and a few everyday essentials. There was no justification for outfitting a large cargo ship, such as the *Trinity*, to an area that could not absorb any goods and could only offer fish and trees. Nonetheless, smaller fishing vessels continued to bring fish to Bristol, a typical cargo consisting of about 9,000 fish.

There are no records of any specific voyages between 1481 and 1490, but several events and records provide evidence that ships ventured to Brassyle throughout the decade.

By 1485, interest may have been cooling. The trade was not profitable, and supplying fish to the Spanish may have become less important as the market matured. The merchants were having trouble justifying the outlay for the small financial return from these voyages.

In 1486, King Henry VII visited Bristol. The new King was anxious to win the support of England's burgeoning merchant class, and he came to talk with the Fellowship of Merchants. Foremost on the agenda were the Mare Oceanum voyages, trade, and the role the new government would play. The Bristol merchants used the occasion to complain to the King that the loss of the Icelandic trade was a financial hardship. The King, however, was not sympathetic and observed that their wives seemed to be arrayed in fashionable and expensive finery.

The King was keenly aware of the voyages to Brassyle authorized by the trading license of 1480-1483. The next step should be to reach Cathay, but the merchants were showing little interest in establishing a foothold in Brassyle or exploring a trade route to the Orient. The King was also acutely concerned about England's relationship with Spain, embodying, as it did, imminent ambitions of expansion.

The King stayed with the Poyntz family on their estate at Iron Acton, a few miles north of Bristol. Richard Amerike was related to the King through his Welsh aristocratic lineage, and he was also distantly related to the Poyntz family through a marriage in 1348.

During the Royal visit, Richard Amerike was appointed as the King's Customs Officer, replacing Thomas Croft.[2] It was a position he would hold several times over the next sixteen years. During his tenure, there was not a single customs record of a shipment of dutiable salt cod imported from Brassyle. If Bristol merchants were importing fish they had caught themselves, it was nondutiable. However, if it was obtained by trading, Amerike may have been a party to concealing this situation from the Crown, or there could have been a tacit agreement with the King so that the merchants could cover the costs incurred in voyages of exploration.

Bartholomew Columbus was now a merchant dealing internationally in charts and books. In 1488, he sailed to London in an attempt to get sponsorship for his brother. The King rejected the Columbus proposal because of commitments he had made to the Bristol merchants.

The King was also in contact with merchant groups in London, and he was watching the developments closely. He made another visit to Bristol in 1490 to discuss these developments and the peace treaty he had reached with the Hanseatic League. The league offered to reopen the Iceland trade to Bristol merchants, but, interestingly enough, the merchants were not interested. They were evidently meeting their salted codfish needs from somewhere else and no longer needed to buy fish from Iceland.

"His companions say that they could bring so many fish that this kingdom would have no further need of Iceland..."

RAIMONDO DI SONCINO

Amerike was reappointed to his Customs position during this visit. King Henry again noticed the "sumptuously appareled" wives of the merchants. This time he extracted £500 from the Mayor and £1 from each man worth over £200.

Adams' Chronicles of Bristol has this entry for 1490:

> "This year divers streets in Bristow were new paved, that is to say, Horse Streat, Knight Smith Streat, Brodestreat, Reclifstreat, St. Thomas Streat, Tuckerstreat, the Backe, St. Mary Port Streat and Lewins Meade, and the High Crosse painted and gilded; the doing whereof cost xxl [£20]. And this year the King and the Lord Chancelour came to Bristow and lay at St. Augustine's. And the commons of Bristow were made to pay King Henry 5 p Cent. for a benevalence."

During this visit, the King may have pressured the Bristol merchants to step up their exploration efforts. It was after all in the national interest to do so. The merchants may have agreed as long as they could be assured that they could at least break even on the costs and not be burdened with taxes on the imported fish.

From 1490, it was common knowledge that several ships ventured across the Mare Oceanum each year.

> "For the last seven years the people of Bristol have equipped two, three, or four caravels to go in search of the Island of Brazil and the Seven Cities."
>
> PEDROS DE AYALA

Spain's Ambassador in London, de Ayala, was reporting to the King and Queen of Spain in 1497 on what was probably the departure of a fishing fleet at the beginning of each season. Spain, in particular, was paying close attention to Bristol ship movements and what was happening in the northern Mare Oceanum.

Thomas Croft died in 1488, William Spencer was now in his 70s, and the other original participants in the Brassyle venture were either retiring or had died. (Spencer would die in 1495 and de la Founte a year later.) A new generation of merchants was stepping into the breach. Richard Amerike was in his 50s and had climbed to a lofty position in local politics as the Customs Officer. Of the original participants in the 1479-1483 expeditions, Amerike was the only prominent member still involved.

Indeed almost all the seafaring community was now drawn from a generation that was familiar with the new fishing grounds in Brassyle. A master mariner named Hugh Elliot and several members of the prosperous Thorne family were younger merchants now working in the Brassyle trade. Hugh Elliot was closely associated with Richard Amerike, and he made multiple crossings to Brassyle in the 1490s. Invariably, he was accompanied by one of the Thorne[3] family. If Amerike was not the owner of one or more of the ships Elliot piloted, he certainly had financial interests in them.

By 1495, the rutters that mariners collected in the Map Room prior to sailing to Brassyle would have had fifteen years of accumulated knowledge. There would have been several copies of this book, duplicated by hand. At least four ships were at times simultaneously making the voyage, probably in convoy. However, each captain would have the full volume on board, and extra copies remained in the Map Room. A mariner of Elliot's stature and competence would have his own copy, which he would embellish himself.

Because of the increase in the number of voyages, Bristol seamen were exploring westward and had sailed far enough along the coast to determine that Brassyle was a mainland—not an island.

> "...It was called the Island of Brazil, and it is assumed and believed to be the mainland that the men from Bristol found."
>
> JOHAN DAY

The usual destinations and major landmarks would certainly have been named by this time. The large body of land to the west, now known to be a mainland, was "Brassyle." But what of the capes, harbors, and settlements they were visiting year after year? Surely the captains and fishermen would have had working names for them by now. Names were not designated as anecdotal whimsy; they identified important landmarks and destinations. Once chosen, these names were written in multiple copies of the rutters so that several pilots traveling in a convoy would have the same information.

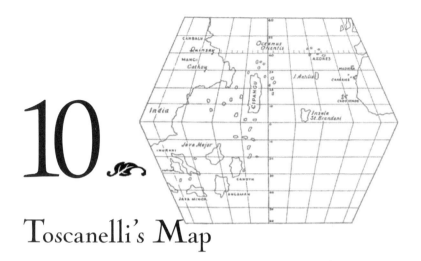

10.

Toscanelli's Map

*"...to Cristobal Colombo, greeting. I perceive your
magnificent and great desire to find a way to where the
spices grow, and in reply to your letter I send you the
copy of another letter which I wrote, some days ago,
to a friend and favourite...of the most serene King of
Portugal...in reply to another which, by direction of his
highness, he wrote to me on [the subject of sailing west
from Portugal to reach the East Indies], and I send you
another sea chart like the one I sent him, by which you
will be satisfied respecting your enquiries."*

LETTER FROM TOSCANELLI TO CHRISTOPHER COLUMBUS
THE JOURNALS OF CHRISTOPHER COLUMBUS,
HAKLUYT SOCIETY, 1893

It had long been presumed that the western edge of Europe and Africa
formed the western edge of the world.[1] Only when it was realized that
the world was a finite globe and the full extent of Asia was known did
cosmographers comprehend that if one traveled far enough to the west,

across Mare Oceanum, one would eventually reach Asia from the opposite direction.

By 1450, European travelers had discovered the continents of Europe, Africa, and Asia, including Cathay and Chipanga, and had produced recognizable maps. These maps, however, were not in general circulation. Ptolemy's theory that the earth was at the center of the universe was still unchallenged, and it would be another seventy-five years before Copernicus would recognize that the earth revolved around the sun. Cosmologists and chart makers, however, were fully aware that the earth was a globe, with cold Polar Regions and a central parallel of latitude called the equator.

The craft of mapmaking was quite sophisticated. Coastlines were charted onboard a ship by triangulating between mountains, headlands, and definitive geographical features. Ships seldom ventured beyond the sight of land, and a captain could accurately chart his position by sighting any three features and calculating the angles between them. Captains did not usually have access to charts before the age of the printing press. However, mapmakers were capturing the tapestries of the coastlines and many were predicting bold new directions.

Paulo dal Pozzo Toscanelli was a brilliant mathematician, astronomer, and cosmographer who lived in Florence where he was a professor at the university. He calculated that the known world of Europe, Africa, and Asia, from Spain to Chipanga, spanned 230 of the 360 degrees of the globe.

Knowing that the distance from Spain to Chipanga was approximately 11,000 miles, he calculated that it would only be a distance of 4,000 miles from Spain to the eastern coast of Chipanga if you sailed west across the Mare Oceanum. He published a map in 1474 showing an uninterrupted Mare Oceanum spanning the remaining 130 degrees of longitude, between Spain and Chipanga.

At the request of King Alfonzo V of Portugal, Canon Fernao Martinez wrote to Toscanelli and requested details of his theory of sailing west to Cathay. Toscanelli replied in a letter dated June 25, 1474, with which he included a map. Columbus also had access to this map, and he would correspond with Toscanelli several years later.

The error in Toscanelli's mathematics would not become apparent for thirty years. The longitude from Spain to Chipanga is approximately 150 degrees, not 230, so completing an encirclement of the earth would entail traveling through 210 degrees, not 130. The distance from Spain to Chipanga, sailing west through the Panama Canal, is therefore closer to 12,000 miles instead of 4,000 miles.

The circumference of the earth at the equator is proportionately greater than Toscanelli thought. He estimated it to be 18,000 miles when in fact it is roughly 25,000 miles.

Ironically, it was Amerigo Vespucci who would discover the true distance. In 1502, he took accurate observations and calculated the circumference of the earth at the equator to within an accuracy of 50 miles.[2]

In 1490, two years before Columbus sailed west, Martin Behaim produced a globe based on Toscanelli's map. It showed Europe, Africa, and Asia, with the Mare Oceanum separating Europe and Chipanga. The world was a small one for the unfolding sphere of exploration and the colorful cast of characters involved. This was the world that Christopher Columbus set out to explore in 1492 when he crossed the Mare Oceanum from Spain to the Caribbean Islands.

Columbus is known to have corresponded with Toscanelli, and he had in his possession a copy of the map. Meanwhile, it is widely believed that Toscanelli may have been a teacher of Amerigo Vespucci when Vespucci was a student in Florence.

Toscanelli was eager to have the Portuguese authorities prove his theory. He had communicated this in 1474 to Canon Fernao Martinez, who had influence with the Portuguese Crown.

Columbus continued to believe that the world was as shown on Martin Behaim's globe up to his death. He had every reason to believe that he had reached islands just to the east of India when he made landfall, and his frustration at not being able to penetrate westward through the landmass can only be imagined.

11.

A Secret Map

Donation "to the ffryres at our lady of Rebedewe to pray ffor us."

FROM THE ACCOUNTS OF THE TRINITY OF BRISTOL
DONATION TO THE FRIARY OF
SANTA MARIA DE LA RABIDA, HUELVA, SPAIN, 1480

After Columbus's return to Lisbon in 1477, his brother joined him, and they both worked as chart makers. They started a small trading business and most likely continued to have contact with Bristol ships and the local agents of Bristol merchants. During his first two years in Lisbon, he traveled frequently to Genoa and Madeira, trading in sugar.

He soon met Felipa Moniz Perestrella, the aristocratic daughter of an Italian diplomat living in Lisbon with his Portuguese wife. Columbus and Felipa married in 1479, and a year later they had a son, Diego. This marriage gave Columbus Portuguese citizenship, which afforded him the right to trade in Africa.

The sea was in his blood, and his attentions turned to the south. Both the Spanish and Portuguese governments were backing exploration along the African coast in the hope that they could open up a trade route around the African continent to India and Cathay.

Portugal got into a conflict with the Spanish or Castillian government over the control of the Cape Verde Islands, which are situated some 300 miles off the west coast of Africa. Castillian forces occupied them between 1477 and 1479, and they began attacking Portuguese ships sailing between Lisbon and the West African country of Guinea. To avoid this threat, ships began to give the Cape Verde Islands a wide berth to the west.

This necessitated being at sea and out of the sight of land for about five days. Such an endeavor required advanced navigation skills based on reading the stars and working with compass bearings. Columbus is believed to have been on at least one of these trips to Guinea and several to Madeira during this time.

Hurricanes form several times each autumn in the sea off the Cape Verde Islands. They travel in a westerly direction to the Caribbean. The tropical ocean currents also follow a similar path. A Portuguese ship returning from West Africa to Lisbon, fortunately laden with food and provisions, is believed to have strayed into one such hurricane and to have found itself several days later at Hispaniola in the Caribbean Islands. The ship eventually made it back to Lisbon, and a rough map of the circuitous route is reputed to have been produced by the captain.

They first went south to the Venezuelan coast and worked their way west and north into the Gulf of Mexico before heading into the Mare Oceanum, seeing the tip of Cuba on the way. Most of the crew died from disease, and the ship was severely damaged from the ravages of storms and teredos.

While Columbus was working in Lisbon in 1479, he met a Portuguese sailor who had survived this voyage and he somehow obtained or prepared a copy of the map. The map showed coastline and islands seen by the sailors. It was not complete but suggested that there were many islands about 750 leagues, or 2,500 miles, west of Spain.

MAP OF SPAIN

Two years later, Columbus corresponded with Toscanelli. He sent Columbus a copy of his map, detailing that Chipanga was 4,000 miles west of Spain, and Cathay was a further 1,500 miles beyond that.

Columbus thought that Toscanelli had overestimated the distance. He believed he would reach the mysterious Island of Seven Cities after about 2,500 miles and that Chipanga would be just another 1,000 miles. This, he believed, was confirmed by the "secret" map he had obtained, and he deduced that he could sail west for 2,500 miles and reach land that was in the vicinity of the Island of Seven Cities.

At age thirty-one, Columbus had sailed as far east in the Mediterranean Sea as Turkey and as far north as the Arctic Circle. He had sailed as far south along the African coast as any European had gone at that time. From his personal exploits, he knew there was a fortune to be made by opening up trade routes to Cathay, Chipanga, and India by sailing west. An ambitious plan to sail west was formulating in his head.

In 1484, he approached King Joao II of Portugal for sponsorship. However, Columbus's demands were quite brash, bordering on rebellious. He demanded one-third of all profits from the trade that resulted and extraordinary recognition for his contribution.

The Royal Court in Portugal did not have much faith in Columbus's scheme. The prevailing winds are onshore in Portugal, and the concept of battling headwinds for 4,000 miles did not seem realistic.

Columbus was still working as a mapmaker in Lisbon in 1484 when his wife died. The political climate, ever sensitive to intrigue, had deteriorated to an alarming degree. The King executed two friends and political acquaintances of his wife's family, the Duke of Braganza in 1483 and the Duke of Viseu (the Queen's brother) in 1484. Late in 1485, Columbus decided to leave Portugal and availing of an opportune connection he moved to Spain.

He stayed at the Franciscan Monastery of Santa Maria de la Rabida at Huelva and, having become close friends with the head Friar, Juan Perez, negotiated to settle his 5-year-old Diego there for the time being. Thus began Columbus's long relationship with the monastery and after that he lived and studied there for long periods of time.

Huelva, the ocean port for the city of Seville, was a center of learning. The monastery welcomed sailors from visiting ships, and many gladly accepted the accommodation that was provided while their ships were laid up in port.

After a few months, Columbus moved to Cadiz, where he was employed by the Duke of Medinaceli. This gave him access to the Royal Court, and in 1487, he began seven years of employment with Spain's royal bureaucracy. He received a small salary, the equivalent of a seaman's wage, and worked as a merchant disposing of war booty. The Spanish had captured Malaga from Moorish control and enslaved the Muslim population.

Columbus returned frequently to la Rabida to visit his son. He spent many hours discussing cosmography with Father Machina, and it was here that his plans to sail to India across the Mare Oceanum were nurtured.

Friar Juan Perez, the head friar, became one of Columbus's strongest advocates.

Merchant ships from Bristol, including the *Trinity*, called at the Friary every few months and Columbus would have been particularly interested in what he could learn about the success the Bristol ships were experiencing in their efforts to locate Brassyle and other islands. Father Machina, who was a friend and confessor to many of the sailors, was undoubtedly a wealth of information.

12.

The Ocean Blue

"And there had beene before that time (1492) a discoverie of some Lands, which they tooke to bee islands, and were indeed the Continent of America, towards the Northwest. And it may be that some Relation of this nature comming afterwards to the knowledge of Columbus, and by him suppressed, (desirous rather to make his Enterprise the Child of his Science and Fortune, then the Follower of a former Discoverie) did give him better assurance, that all was not Sea, from the west of Europe and Africke unto Asia."

THE HISTORY OF THE REIGN OF
KING HENRIE THE SEVENTH. FRANCIS BACON, 1622

A fter the English crown rejected Columbus's plea for sponsorship, he again plied his quest to the Queen of Spain, but without success. In 1491, he discussed this with Friar Juan at Huelva, where his son still lived. Soon after, the Abbot met with the Queen, and he advocated for Columbus. He told her that Columbus was ready to leave

Spain and go to England where he felt sure he could convince King Henry VII to support him. The Queen told the Abbot that when Spain had finally expelled the Moors from Castile and drove them back to North Africa, she would consider his plea with commitment.

Granada, the Moorish capital, fell in February 1491. True to her word, Queen Isabella arranged financing, and she agreed to most of Columbus's exorbitant terms.

Columbus had his ships provisioned by the Medici Bank in Seville, where a young banker named Amerigo Vespucci made appropriate arrangements.

On August 3, 1492, Columbus left the Spanish port of Palos with three ships: the *Nina*, the *Pinta*, and the *Santa Maria*. They sailed southwest to the Canary Islands from where they turned and headed due west across the Mare Oceanum. Columbus knew that the trade winds in the latitudes of the Canaries would be astern, blowing him south and west into the unknown world beyond the Atlas.

During the voyage, Columbus privately referred to his secret map, from which he had calculated the length of the sea crossing and the time it would take. The *Santa Maria* was navigated by its owner, Juan de la Cosa, who would later publish a map that is still controversial.

On October 12, they sighted land and arrived on the island of San Salvador in the Caribbean. Columbus was convinced that they had reached some islands that were just to the east of India. He immediately referred to the location as the "Indies," and he called the inhabitants "Indians." The *Santa Maria* ran aground on a reef and was broken up by the surf. They worked their way north in the remaining two ships through the islands, reaching Cuba where they looked for gold, with no success, before returning to Spain.

Columbus arrived back in Spain on March 15, 1493, stepping ashore in Palos, near Huelva. The King and Queen were at the other end of the country in Barcelona, and Columbus, enlivened by his success, seized the opportunity to parade his exotic cavalcade. He immediately set out overland to report to them. This trip took him through many towns,

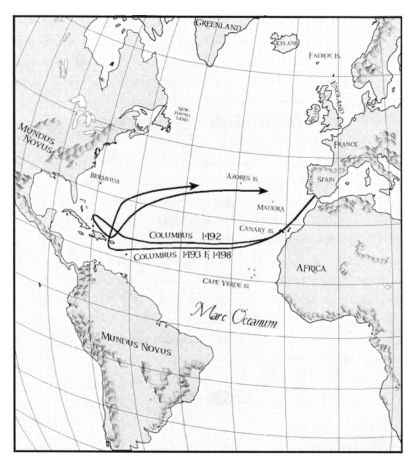

COLUMBUS'S VOYAGES TO THE NEW WORLD

including the port city of Valencia, where his entourage, complete with parrots, wondrous plants, and Indians, paraded through the streets. (A certain John Caboto witnessed this outlandish cavalcade. Realizing that Spain or Portugal would no longer back his proposal, Caboto would turn to the English for help in his quest to find a passage over the Mare Oceanum to Cathay.) When he arrived in Barcelona, Columbus reported to the authorities that he had reached islands off the coast of India, and he held this belief until the day he died in 1506.

Spain laid claim to these new discoveries, and the Pope supported this position. At the Treaty of Tordesillas in 1494, a formula was devised where the New Lands were to be divided between Spain and Portugal. Portugal claimed all non-Christian lands east of a line of longitude drawn 370 leagues to the west of the Cape Verde Islands. This effectively gave most of the New World to Spain. Brazil and Africa fell to Portugal.

Motivated by the prospects of wealth, Columbus immediately planned another expedition. This one would be larger and better equipped. On September 25, 1493, he sailed again with 17 ships, 1,500 men, and livestock. They established a community on Hispaniola and visited Santo Domingo and Puerto Rico. Columbus returned to Spain nearly three years later on June 11, 1496, with 500 slaves, gold, and tobacco.

On his first voyage to the Caribbean in 1492, Columbus seemed to know exactly where he was going, and it seemed to his crew and associates that he was continually attempting to fit his discoveries into some preconceived map. On subsequent visits, he consistently refused to accept that Cuba was an island, and forbade anyone to sail west along the southern shore to prove to the contrary.

13.

Cabot's Voyages

"Henry, by the grace of God, king of England and
France, and lord of Ireland,...
Be it knowen that we have given and granted...to our
well beloved John Cabot citizen of Venice, full and free
authority to saile to all parts, countreys, and seas of
the East, of the West, and of the North, under our
banners and ensigns,...and have given license to set up
our...banners and ensigns in any town, city, castle, island
or mainland whatsoever, newly found by them..."

TRADING LICENSE GRANTED TO JOHN CABOT,
KING HENRY VII, 5 MARCH 1496

John Cabot, another experienced Italian mariner, was living in Valencia in 1493. In April of that year, Columbus passed through Valencia after his first voyage to the Caribbean en route from Seville to Barcelona. Cabot spent part of his early childhood in Genoa, and he and Columbus may have known each other. John grew up on the streets and canals of Venice, where his family moved in 1461. His father was a merchant.

Like Columbus, Cabot had ambitions to sail to the Orient across the Mare Oceanum. He was already well traveled and had been as far east as Mecca. He correctly believed that it would be a shorter distance to Asia at a more northerly latitude, and he tried to persuade Spanish and Portuguese merchants to sponsor him in exploring a more northerly route than Columbus had taken. He was unsuccessful.

Cabot realized that neither Spain nor Portugal would be likely to sponsor him after Queen Isabella sided with Columbus. He decided to approach the English government, reasoning that this would afford an opportunity for England to catch up with Spain and Portugal without getting into a competitive debacle in the southerly latitudes.

Cabot had also heard that Bristol merchant ships were regularly crossing the open seas to the island of Brassyle, which he calculated to be near the northeast coast of Cathay. He was using the same Behaim globe that Columbus had relied on, and he expected to sail about 1,500 miles from England and arrive at the coast of Cathay by using this northerly latitude.

Cabot thought that the Bristol merchants might be persuaded to take him to Brassyle on an officially sanctioned expedition. At worst, he thought their fishing base could provide a safe harbor and staging area for further exploration.

He approached a relative of a Bristol merchant in Valencia and through this intermediary the Fellowship of Merchants agreed to sponsor him. Cabot and his family sailed to Bristol early in 1495.

No doubt the King and the English government realized the potential for controlling a northern trade route to Cathay. And by cunning and diplomacy, they might be able to claim new territory without competing with Spain. The King was particularly anxious not to antagonize Spain. He was negotiating a marriage between his son Arthur and Catherine of Aragon of the Spanish court. They married in 1501, but Arthur died the following year. When Henry VIII inherited the throne in 1509, he

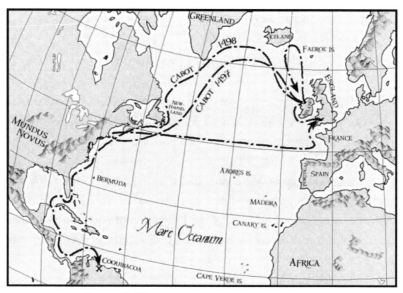

CABOT'S VOYAGES TO THE NEW WORLD

married his deceased brother's widow, who became the first of his six wives.

The Bristol merchants appeared focused only in importing fish and immediate trade opportunities. They had failed to find any Cathay communities with which to trade during their explorations along the west coast. It seemed to them a lost cause.

The Crown was not unduly concerned about having a foreigner in control. The Bristol merchants with their trading license had long since decided that there was no pressing reason to claim Brassyle for England.

Cabot's goal was primarily to open a trade route to Cathay. Any land claims would be secondary and could be important strategically to control the lucrative trade routes. Cabot, like Columbus, was driven by the promise of personal wealth that would derive from opening up trade routes to the spice and silk markets in the Orient.

> *"By this means they hope to make London a more important mart for spices than Alexandria."*

RAIMONDO DI SONCINO

Cabot obtained the necessary trading license, which was issued at an audience with King Henry VII on March 5, 1496. It stipulated that he could take five ships, he must stay to the north, and he must go from and return to Bristol.

The Fellowship of Merchants had the connections in London and probably made the application for the license as they had in 1479. Richard Amerike was a senior member of the Fellowship of Merchants and had been the Customs Officer for ten years. He was in the perfect position to make the arrangements.

An expedition like this was expensive, so only one ship was put at Cabot's disposal. The *Matthew* was a 3-masted 50-tun caravel measuring about seventy feet in length. Amerike was the largest investor in the *Matthew* voyages.

Cabot was disappointed that the Bristol merchants would not provide the security of at least a second ship. They did not sail until the autumn of 1496, which was late in the season. Other ships of the Bristol fishing fleet had already gone to Brassyle and had not returned yet. The ship left Bristol and headed over the well-traveled route to Ireland, past the Faeroes, and on towards Iceland. The prevailing winds at these latitudes are from the southwest. Sailing north keeps land in sight and avoids bucking head winds.

Cabot traveled with his barber, the only person on board with whom he could communicate. He spoke Genoan Italian and Spanish, and the Bristol men spoke in a strong regional English dialect. Even an Englishman from London would have had a difficult time understanding them.

Relations on board the ship were strained. The weather was severe; crashing storms and ice-cold rain beat down on the crew. They were concerned about ice building up in the rigging and the possibility of the

86

ship capsizing. Unused to sailing in icy conditions, Cabot may not have appreciated the danger. Cabot was a skilled mariner who was well schooled in all kinds of navigation and probably wanted to navigate by the stars. The merchants may have intentionally confused Cabot for misguided or xenophobic reasons. They probably did not want a foreigner learning too much about their business.

For whatever reasons, communication and teamwork were at a low ebb, and the ship did not even get as far as Iceland. When the weather turned violent, Cabot gave the order to turn back to Bristol.

> *"Since your Lordship wants information relating to the first voyage, here is what happened: he went with one ship, his crew confused him, he was short of supplies and he ran into bad weather, and he decided to turn back."*
>
> JOHAN DAY

✛

A few months later, in the spring of 1497, they would try a second time. Hugh Elliot, the acclaimed mariner and himself a merchant, would be the master of the ship. Robert Thorne and another Bristol merchant would accompany him. Elliot and Thorne were both in their thirties. That these well-traveled and well-financed merchants from Bristol would embark on a potentially dangerous trans-oceanic voyage with one small 70-foot vessel shows a high level of confidence in their ships, their sailing abilities, and their knowledge of where they were going. They had authorization to take five ships, and a prudent mariner would take at least two or three for contingencies in case disaster struck. But it also may indicate that the merchants were mindful of the pecuniary nature of such a trip: the lack of profits involved in exploring a land they knew was not the promised spice kingdom. During these exploratory missions, they purposely avoided their own fishing grounds. Under these circumstances, there would be no justification for sending more than one ship or even a larger vessel.

The ship was loaded with thirty-five barrels of salt and fifteen barrels of beer and wine. Drinking water, stored on board in barrels, would usually be undrinkable after a week. Barrels of dried beans, dried peas, salted meats, fish, vegetables, small pigs, some chickens, and enough food for seven months were loaded into the hold.

On May 2, 1497, the *Matthew* sailed from Bristol's Redcliffe Wharf. Twelve Bristol crewmen sailed the ship. Cabot again took his barber and a merchant from Burgundy. Other ships of the Bristol fishing fleet had already left for Brassyle, and his conspiring crew, made up of local seamen, knew that it was important that Cabot did not see them or know where they were going. They went via Bantry Bay on the west coast of Ireland on the same route to the fishing grounds of Iceland, then due west, skirting the southern coast of Greenland.

Again, there were conflicts between the Bristol crew and Cabot. The *Matthew* sailed south of the fishing grounds off the coast of Newfoundland. This may have been by design if Elliot and Thorne did not want Cabot to observe the fishing area.

On the other hand, Cabot, with his navigation skills and knowledge of Toscanelli's map, may have been confident that they could stay to the south in the open sea and approach the mainland of Brassyle further to the west.

The crossing from Ireland was relatively easy with only one gale. After thirty-four days, they sighted land in the vicinity of Maine, or Cape Breton. The coast was heavily forested with tall straight trees, and the weather was clear. This suggests that they were considerably west of Newfoundland, which is more lightly forested and usually foggy in the summer. They noted that the trees would be a good source of timber from which to make ship masts.

> "…and they found tall trees of the kind masts are made, and other smaller trees, and the country is very rich in grass."
>
> JOHAN DAY

REPLICA OF THE MATTHEW

The Bristol merchants knew exactly where they were. They saw evidence of fish drying on racks on the beaches and other evidence of human habitation. Cabot cautiously went ashore on June 24, staying within gun shot range of the ship in case there were hostile natives. He planted the Royal Ensign and claimed the land for England.

Elliot had the reputation of being the "leading pilot in England." He and Thorne had visited these shores together at least once before. Elliot had been issued his rutters in Bristol, and he may also have had his own rudimentary charts. Cabot no doubt was anxious to see and copy them. The Bristol men would also have their own names for many places, and these names may have been in Elliot's records.

The trip had to be paid for, so the crew fished and salted cod to take back to Bristol. The cargo of fish was valued at £200 on their return to Bristol. Afterwards, Cabot wrote about the fish to his friend Soncino, who in turn quoted Cabot in another letter.

> "The sea is swarming with fish, which can be taken not only with the net, but in baskets let down with a stone...his companions say that they could bring so many fish that this kingdom would have no further need of Iceland, from which place there comes a great quantity of stock-fish."

RAIMONDO DI SONCINO

The *Matthew* then traveled east along the coast for a month, past Nova Scotia, New Brunswick, and the southern coast of Newfoundland. Cabot drew extensive charts of about 900 miles of coastline. He was convinced that he had reached an island or an extension of land off the Cathay mainland.

On his own map, Cabot named the initial landing area "Prima Vista," and he identified the mainland as "Brassyle." A nearby island was named "St. John's Island" because June 24th is St. John's Day. He thought the large island to the east (which may have been New Brunswick) must be the Island of Seven Cities. He named several capes and landmarks, some of them after friends and acquaintances. The name "Cape St. George" survives to this day and may have originated from this voyage.

After leaving the coast, they sailed back to England, arriving in Bristol only fifteen days later on August 6, 1497.

Just as they had done on the outward voyage, the crew misled Cabot on the homeward crossing by swinging further to the south than their usual route. Confident that they would reach land in France, they did not worry that they were well off course. When they sighted the coast of Brittany, they established their location and sailed north to England.

After returning to Bristol, Cabot and the merchants immediately went to London. Just three days after docking in Bristol, they were granted an audience with King Henry VII. A meeting was held at which the Spanish envoy, Pedros de Ayala, attended, and Cabot brought his map and a globe showing the part of Cathay they had visited and how near they were to the spice and silk markets. The King renamed the new territory "New founde land" at this meeting, and he awarded Cabot a pension of £10 for that year. This was subsequently raised to £20 a year on December 13, 1497, of the same year.

Cabot's contribution to the Brassyle venture had been to identify Brassyle in the global picture. It was surely a part of northeast Cathay, and it would just be a short journey down the coast to the large cities of Cathay. The silk and spice markets were almost within reach.

After his audience with the King, Cabot was feted as a hero, dressed in silk, and called "the Great Admiral." When the party returned to Bristol, other names were added to the map. In a state of euphoria, he celebrated by bestowing titles and land to friends and naming landmarks in the newly recognized land. Many believe that Cabot named an island or territory after his sponsor, Richard Amerike. Raimondo di Soncino, Milan's representative in London and a friend of Cabot, wrote to the Duke of Milan in December 1497:

> "The Admiral, as Master Johan (Cabot) is styled, has given a companion an island, and has also given another to his Barber, a Genoese. Some Italian friars have the promise of being bishops..."

RAIMONDO DI SONCINO

Cabot's map does not exist today, so nobody knows for certain which names he added. (See Appendix E for a colorful newspaper article printed in Bristol in 1943.) Everyone was convinced that Cabot and his party had visited the east coast of Cathay. No one gave thought that a new continent had been discovered. It would be six years before this belief would be challenged.

King Henry was so enthused about Cabot's discovery that a third, far better equipped voyage was immediately planned for the next year. Another trading license was issued on February 3, 1498, authorizing Cabot to take 6 ships, each of up to 200 tuns, and to explore the coastline south from Newfoundland to where the Spanish were concentrating their activities.

Cabot's map from his 1497 voyage was seen and copied by several people. To make their land claim, the English government had to allow representatives of foreign powers to see it. Pedros de Ayala wrote a detailed report to his government in July 1498. (See Appendix A.)

Columbus and Vespucci did not have to wait for the Spanish government to report to them on what Ayala had seen on Cabot's map. As incredible as it sounds, a merchant in Bristol sent Columbus a copy of Cabot's map. (See Appendix B.) This discovery was made in 1955:

> "I am sending you a copy (map) of the land which has been found."

> JOHAN DAY

Columbus had possession of Cabot's map before he sailed on his third voyage to the Caribbean in 1498, and Amerigo Vespucci had until the spring of 1499 to study it. (Additional letters describing Cabot's discoveries in the New World have surfaced over the years. (See Appendix C.)

14.

Columbus and Cabot in 1498

*"It is certain that Hojeda in his first voyage encountered
certain Englishmen in the vicinity of Coquibacoa"*

MARTIN FERNANDEZ DE NAVARETTE,
SPANISH HISTORIAN, 1829

In May 1498, both Columbus and Cabot coincidentally sailed west to
the New World, both intent on completing their sea routes to India
and Cathay. Columbus sailed from Spain with six ships, and Cabot
sailed from Bristol with five ships.

Cabot's fleet consisted of one large ship provided by the King at a cost
of £113-8s-0d and furnished by Lancelot Thirkill, a London merchant.
Bristol traders provided four smaller merchant ships that were loaded with
"slight and gross merchandises, as coarse cloth caps, laces, points, and other
trifles, to trade with Indians." There were 300 men on board the ships,
signifying the size and importance of this expedition. The cargo indicated
that they expected to meet and trade with Indians, an assumption that
Cabot himself would not have made based on his own experience the
previous year.

A few days after leaving Bristol, one of the ships was damaged by a storm off the Irish coast, and it eventually returned to England. The citizens of Bristol expected Cabot's fleet to return within two years, but nothing was ever heard from them again. The entire fleet disappeared without a trace. However, they did leave a few clues.

The remaining four ships sailed the usual northern route, skirting Greenland and reaching the coast of the New World in the vicinity of Labrador and the northeast coast of Newfoundland. Cabot had not been this far north in 1497. They then sailed south along the Atlantic coast of Newfoundland. At Grates Cove, near the southeast tip of Newfoundland, one of the ships struck a rock and broke up. The crew made it to shore.

Five years later, Gaspar Corte Real, a Portuguese navigator, rounded up fifty native Beothuks at Grates Cove and took them prisoner to be sold as slaves. One had in his possession a broken Venetian sword and another had an Italian earring.[1]

Cabot was not at the mercy of the Bristol merchants on this expedition. If there were fishing boats, he would have seen them. The place where Amerike's salt had first been delivered sixteen years earlier might have been pointed out to him by crew members who had been there before. Perhaps they visited a familiar safe harbor, and Cabot may have included it and its name on his charts.

The fleet, or the three ships that now remained, rounded Newfoundland and sailed west passing the coast they had surveyed the previous year. They continued west and south charting the present day United States Eastern Seaboard, all the time looking for signs of Cathay. They landed and went ashore in several places. They rounded the southern tip of what is today Florida, and sailed west into a large gulf following the coastline around to the south.

In just twelve months, the concept of Brassyle being an island changed. It was evidently part of a very large landmass. Perhaps the Bristol merchants had been right after all when they had abandoned notions of finding the cities of Cathay. By August 1499, they were sailing east along the north coast of what is now Venezuela.

✟

In May 1499, four caravels set sail from Cadiz, Spain, under the command of the 29-year-old Spanish Admiral Alonzo Hojeda. Second in command was Juan de la Cosa, who had sailed and owned the *Santa Maria* in Columbus's first expedition in 1492. He was now nearly fifty years old. Also on board was Amerigo Vespucci.

Hojeda was by all accounts a ruthless and cruel leader, and his behavior during this voyage would prove no exception. One of his ships was not to his liking, so Hojeda put into port and appropriated another ship by force, leaving the first ship in exchange. He tricked several ships into letting himself and a band of his men board from a rowboat, and once on board, they robbed the ships. In the Canary Islands, they looted the house of Dona Ines de Peranza, the daughter of Columbus's mistress, Dona Beatriz Enriquez de Arana.

Hojeda crossed the Mare Oceanum and arrived on the New World coast near the mouth of the River Orinoco and then sailed westward along the coast robbing, fighting, and killing any natives they encountered. They burned one settlement to the ground. In the Gulf of Maracaibo, they fell upon a coastal village built on stilts in the swampland. They robbed the villagers of their gold in a house-to-house search, killed a score of the native men, and kidnapped a young girl.

When they reached Coquibacoa in August, they met what was left of Cabot's fleet, which was sailing east towards them.

Malaria, disease, and teredos had taken their toll. What happened to Cabot's men was not documented, but evidence indicates that they were slaughtered by Hojeda's forces.

Admiral Hojeda's encounter was the last time anyone saw them alive. Cabot was trespassing on land that had been predetermined as Spanish by Papal edict. Hojeda wanted to prevent England from making any claims that would conflict with Spanish ambitions.

A Spanish historian, Martin Fernandez de Navarette, wrote in 1829:

> It is certain that Hojeda in his first voyage encountered certain Englishmen in the vicinity of Coquibacoa.

Other than the Cabot expedition, there were no other English expeditions in that area at that time. Navarette's source is unknown, but he was a widely respected historian in his day.

If this is what happened, Amerigo Vespucci and Juan de la Cosa may have been present at this ghastly deed, and Cabot's maps could have been taken in the encounter. This would have been the second time that Cabot's extraordinary efforts to produce a map of the New World would wind up in the hands of the Spanish.

From Venezuela, Hojeda and de la Cosa sailed north with two of the ships and joined Columbus at the settlement he had established at Hispaniola. The ships had to be laid up to repair damage they had suffered, some say resulting from the battle with Cabot.

✠

Vespucci, an Italian gentleman, may have had enough of the violence he had been obliged to witness. He left Hojeda and sailed with a second ship southeast along the north coast of what is now Brazil. He believed he was traveling along a peninsula of the Asian continent and would eventually round the southernmost point and be able to sail west to India.

Vespucci happened to have with him some astronomy charts for the Italian city of Ferrara. He was becalmed for almost a month at a point along the Brazilian coast, and he used this time to calculate his position in longitude by comparing the movements of the planet Mars with where it would be if he were in Ferrara. According to the charts, on August 23, 1499, Mars was projected to be in conjunction with the moon at approximately fifteen minutes before midnight. Where Vespucci was positioned, it had occurred several hours earlier. In fact, it had occurred before the moon had risen, which was ninety minutes after sunset. Using his observations, he was able to accurately measure the time difference and mathematically calculate his distance from Italy. This was how he discovered the key to calculating the exact size of the terrestrial globe.

✠

Juan de la Cosa returned to Spain with Hojeda, arriving in the port of Santa Maria in July 1500. De la Cosa drew his map of the Caribbean and the New World while his ship was moored in the harbor there. It shows details of English landings between what has been interpreted as Labrador in the north to as far south as South Carolina.

De la Cosa must have used Cabot's charts to prepare his map. The coastline west of Coquibacoa is drawn with surprising accuracy, in spite of the fact that de la Cosa had not ventured that far west. Indeed the first map shows a definitive coastline around Florida and the north shore of the Gulf of Mexico. Officially, no European visited that area until 1513. This map is the earliest surviving map of the area. Written in the corner is:

"This map was made by Juan de la Cosa
at Puerto de Santa Maria in 1500."

Juan de la Cosa's map, which is now displayed in the Naval Museum in Madrid, shows the east coast of the New World with English flags in five locations along with the words "mar descubierto par inglese" (sea discovered by the English). Most likely, the English flags represented Cabot's landings. This map also shows Cuba as an island. Columbus had insisted it was connected to the mainland.

Vespucci continued southeast along the Brazilian coast, and he is credited with discovering the mouth of the River Amazon. His ships eventually crossed the equator and reached Cape St. Augustine, which is at the latitude six degrees south. Here the coast turns south-southwest, and he abandoned his voyage further south because his ships became riddled with teredos. He turned back, retracing the same route as far as Trinidad and then north to Haiti. He then headed back to Spain and arrived in the summer of 1500 at the port of Santa Maria, where his friend de la Cosa was working on his map. Vespucci did some writing and worked on his own maps, also having access to the Cabot charts. De la Cosa was preparing for his next exploration, and he left port in October 1500 and returned to the area of Columbia in the Caribbean.[2]

✠

Meanwhile in Bristol, another expedition was preparing to go to Brassyle. In March 1501, two Bristol merchants teamed up with three mariners from the Azores for a voyage to Labrador. Thomas Ashhurst, John Thomas, Joao Fernandes, Francisco Fernandes, and Joao Gonsalves sailed from Bristol. They explored the Labrador coast and made another attempt at finding a passage to Cathay around the New founde land. They returned to Bristol late that year, though Joao Fernandes died during the voyage.

The following year, Hugh Elliot, the pilot on the *Matthew* in 1497, joined Fernandes, Ashhurst, and Thomas for yet another attempt. They returned to England with three natives, who lived in London for several years.

Even at this time there was still no suggestion that the land that was attracting so much attention was a new continent. The English, the Portuguese, and the Spanish still believed they were exploring peninsulas and islands adjacent to Cathay and India. It was obvious, however, that the place they earlier called the island of Brassyle was a landmass.[3]

The voyages that originated from Bristol were still going to the same general area that they had first visited twenty-two years earlier and were still piloted by some of the same people. The rutters that were issued to Elliott and other crews, as late as 1503, were still derived from the same names that had been in use down through the years.

In Bristol, some merchants formed "The Company of Adventurers to the New founde land." The purpose was to import fish to Bristol. In 1503, a shipment was brought back. While it was dutiable, the importers were able to get an exemption. It was recorded in the customs records, but no duty was paid. Richard Amerike had recently died, and perhaps the new Customs Officer did not feel comfortable continuing with the "old arrangement," if Amerike had in fact been turning a blind eye to previous imports that had been subject to duty.

✠

Admiral Hojeda, far from being condemned for his atrocities, was actually rewarded by the Spanish Crown for his patriotic work in stopping the English encroachment from the north. He was awarded the title "Governor of Coquibacoa" and was given instructions to proceed with his discoveries:

> "that you go and follow the coast which you have discovered, which runs east and west, as it appears, because it goes towards the region where it has been learned that the English were making discoveries…in order that it be known that you have discovered that land, so that you may stop the exploration of the English in that direction."

Further reference to Hojeda's battle with Cabot is contained in a license awarded to him at a later date:

> "Likewise their Majesties make you a gift in the island of Hispaniola of six leagues of land…for the stopping of the English, and the six leagues of land shall be yours forever."

The news of any such battle, however, would never reach Bristol, and the fate of Cabot would remain a mystery.

15.

Quarta Orbis Pars

"we may rightly call (it) a New World...
it transcends the view held by the ancients...
that there was no continent to the south
beyond the equator.
In those southern parts I have found a continent
more densely peopled and abounding in animals
than our Europe or Asia or Africa."

MUNDUS NOVUS, AMERIGO VESPUCCI, AUGUST 1504

After Vespucci returned to Spain in the summer of 1500, he made plans to resume his search for the southernmost point of the Cathay peninsula and then sail to India. He would travel further south along the east coast of South America starting at Cape St. Augustine, where he had left off before. Because he would be sailing along land the Pope had granted to Portugal, he decided to approach the Royal Family in Lisbon for sponsorship, which was granted.

He sailed from Lisbon on May 14, 1501, first to the Cape Verde Islands, 300 miles off the West African coast. He then sailed southwest across the Mare Oceanum to Cape St. Augustine where he turned south.

He reached the site of the present day Rio de Janeiro in January 1502 and continued a short distance past the mouth of the River Plate, which he is also credited with discovering. He charted the stars in the southern sky and tried to determine the star that was situated over the South Pole. This would constitute a remarkable scientific breakthrough for navigators sailing in the southern hemisphere.

Vespucci sailed nearly 3,000 miles along the new coast trying to reach the end of the peninsula before giving up. He returned to Portugal via the island of South Georgia in the south Mare Oceanum, arriving in Lisbon in the summer of 1502.

It was during this, his "third" voyage that he finally came to the realization that the new coast was not part of Cathay. It was, in fact, a fourth continent. This enlightened development, together with his emerging techniques of measuring the globe, would fundamentally change the map of the world.

After his return to Portugal, he wrote a short book titled *Mundus Novus* (New World), which described the plants and wildlife he saw along the coast and the native peoples he met. His book gained notoriety because of his lurid and promiscuous descriptions of the native women. In one incident, a sailor was seduced by a group of women and then killed. His flesh was cooked and eaten in full view of the ship. The crew was too frightened to intervene. *Mundus Novus* was a "bestseller" and was published in fourteen editions and in several languages.

Vespucci had a long correspondence with Piero Soderini, a lifelong friend who was now the Gonfaloniere, or Head Magistrate, of Florence. He sent him detailed letters of his voyages, one of which described his method of calculating longitude. He showed how he calculated the earth's circumference and came to the conclusion that the new discoveries constituted a separate continent. These letters were published in at least ten editions.

He also wrote detailed letters to the Medicis. It is the apparent conflicts between these three different manuscripts that have contributed

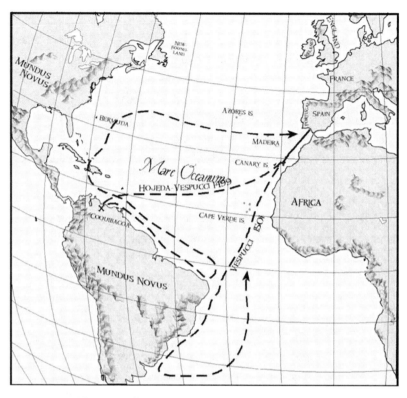

VESPUCCI'S VOYAGES TO THE NEW WORLD

to the disputes about the first and fourth voyages that have been attributed to Vespucci.

Vespucci most likely sent copies of the Soderini letters and maps to a young scholar, Matthias Ringmann, who was working at a well-known academic institution near Strasbourg. Like many other facets of this story, the Ringmann-Vespucci connection would have far-reaching consequences. The letters were written in Latin and Ringmann translated them. An edition of *The Three Voyages of Amerigo Vespucci*, which included a poem written by Ringmann, was published in Strasbourg. The three voyages described were those of 1497-1498, 1499-1500 and 1501-1502.

Some historians doubt that Vespucci made the first voyage. There is irrefutable evidence that he was in Florence in 1497. If he had sailed to South America in 1497, he, not Columbus, would have been the first European to set foot on the American mainland. Some people believe that Vespucci intentionally fabricated this version; others think he had no hand in creating the deception. (See Appendix D for excerpts from Vespucci's letter to Soderini on the disputed first voyage.)

He may have made a second voyage bearing the Portuguese flag in 1505 under the command of Gonzalo Cuelho with Juan de la Cosa navigating a second vessel. However, this voyage did not produce any discoveries of importance, and some doubt that Vespucci participated. After this voyage, he sent more letters to Soderini, and the publication was updated. *The Four Voyages of Amerigo Vespucci* was again widely published in many editions and languages. This document would become the basis for a World Map that was printed in St. Die, a remote village hidden away in the Franco-German borderland, far from either Lisbon or the Mare Oceanum.

In 1505, Vespucci was invited to Spain for private consultations where he was offered a high-ranking position with the Casa de Contratacion de las Indias (the Commercial House for the West Indies). He moved back to Spain, married Maria Cerezo, and was granted Spanish citizenship. In 1508, he was appointed pilot major, or chief navigator, of the state. In his capacity as the head of the Spanish Admiralty, he trained and granted licenses to pilots, and he controlled the master map for the entire Fourth Continental area.

Vespucci died in Seville in 1512 as a result of complications he developed from malaria, a disease he had contracted in the "mundus novus."

Vespucci has been discredited, particularly in the last 200 years, because of the discrepancies between the contemporary accounts of his various voyages. However, the fact that the Spanish government would appoint him, a foreigner, to one of the most important positions in Spain, is ample testament to his reputation and abilities.

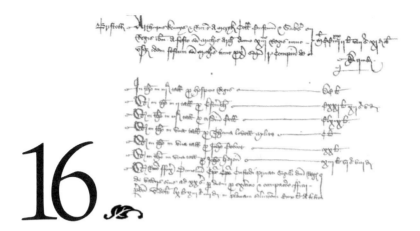

16

Discovery of a Spanish Filing Error

*"Bristoll Arthurus Kemys et Ricardus ap Meryke
Collectores Custumarum et Subsidiorum Regis ibidem
a festo Sci. Michaelis archangeli anno xiii Regis nunc
vsque idem festum Sci. Michaelis tunc proxime sequens,
reddunt Computum de
mccccxxiiii. li. vii. s. x. d. q.
De quibus*

Et in thesaurario in   *una tallia
pro Johanne Cabot*   *xx. li"*

CUSTOMS ROLLS, BRISTOL, 1499

In 1892, the United States of America celebrated in grand style
the 400th anniversary of Columbus discovering America. Five years
later, the authorities in the city of Bristol observed the 400th anniversary
of the Cabot voyage of 1497 to bring attention to his discovery of the
North American mainland. They built a tower in the city and staged other
special events.

105

This revived interest led to the publication of several books on the subject, which brought to light an immense amount of information about Cabot that had not been known before. Nothing was known up to that time of the fate of Cabot after he left Bristol on his second voyage in 1498. Many people wondered if he had in fact returned to England. The complete lack of information dating after 1498 led historians to conclude that he had perished at sea.

Earlier in the 1890s, Mr. Edward Scott, a historian working at the British Museum, discovered some English government Exchequer financial documents in the Chapter House of Westminster Abbey. Two of these ledgers were the customs and excise collections and expenditures for the City of Bristol for the years 1498 and 1499. On both, a payment to John Cabot of £20 for his pension was recorded. It was later thought that Cabot had returned to Bristol after his second expedition (1498) because his pension had been paid. It now appears that it was paid to his wife. Mr. Scott contacted a colleague in Bristol who was intrigued, and the documents were taken to Bristol and examined. A short time after, they were published.

Mr. Alfred Hudd, a local historian in Bristol, came across them. He noticed that a name on each document identifying one of the Customs Officers was uncannily close to the name "America" in each case.

On one ledger, the name was Richard Ameryk, and on the other, it was Richard ap Meryke. It transpired that Ameryk and ap Meryke were the same person.

Translated from the Latin, the document stated in part:

"Bristol.
Arthur Kemys and Richard ap Meryke collectors of the King's customs and subsidies there, from Michaelmas in the 14th year of this King's Reign till the same feast next following, render their account of £1,424 - 7s - 10 1/4d

Out of which
In the treasury in one tally
in the name of John Cabot £20 - 0s - 0d"

This discovery inspired further research about this man Amerike. He was connected with the brass plate on the tomb of John and Johanna Broke in St. Mary Redcliffe Church and was identified as Johanna's father. His name also appeared in civic records in connection with his roles as the King's Customs Officer and later as the Sheriff of Bristol, and he was known to be a member of the Fellowship of Merchants.

Arthur Kemys, the other name mentioned, was Amerike's assistant at the Customs House. He came from a prominent local family. A rent ledger was discovered showing that John Cabot's wife rented a house owned by Kemys for £2 a year.

This was enough for Alfred Hudd to conclude that Cabot may have named America after Richard Amerike. Alfred Hudd first put forth his theory about the derivation of the name "America" from Richard Amerike to the Clifton Antiquarian Society in a lecture in 1908. Since that time, more evidence has come to light showing Amerike's involvement in trade and his associations with the participants in the trading license explorations in the years 1480 to 1483, especially with Thomas Croft, his predecessor at the Customs House.

✚

After 40 years, more evidence came to light. A letter discovered in 1955 in the Spanish National Archives had far-reaching implications for Bristol. It established that Bristol merchant ships had sailed to America considerably earlier than Columbus had, something that had long been suspected but was difficult to prove.

Dr. Hayward Keniston, a Romantic language expert from the University of Michigan, was researching old letters in the Spanish National Archives building in Simancas, Spain. He discovered a letter written to

Christopher Columbus by Johan Day, a Bristol merchant. It was written between Christmas 1497 and March 1498. In Spanish, it contained a long report of the Cabot voyages of 1496 and 1497 and was clearly one of a regular correspondence between Columbus and Day. The following is an excerpt of the full transcript provided in Appendix B: [1]

> *"It is considered certain that the cape of the said land was found and discovered in the past by the men from Bristol who found "Brazil" as your Lordship well knows. It was called the Island of Brazil, and it is assumed and believed to be the mainland that the men from Bristol found."*

Johan Day, 1497

Dr. Keniston mentioned the letter to a colleague, Dr. Louis Andre Vigneras, who immediately realized its importance. The letter had not been noticed before because of a simple clerical error. It had been filed with documents pertaining to English voyages to Brazil, when in fact it referred to Brassyle. The name "Brazil" referred to the vast tract of new land discovered south of the equator in the New World. It was named for the hard brazil wood that was found to flourish there. The voyage in question, however, referred to the mythical island that was purported to exist off the west coast of Ireland and was known by its Celtic name "Brassyle." The unearthing of this important letter introduced yet another colorful character to the fifteenth century stage.

Johan Day had lived in Andalucia in southern Spain, and he may have been working for Spanish interests in Bristol. The letter was written by Day while he was in the Spanish port of Santa Maria, near Cadiz. He could have been a spy, but whatever his role, his letter is one of the better accounts of Cabot's voyages.

This letter completely changed the understanding of the roles that Columbus, Cabot, and the Bristol merchant traders played in the discovery of North America between 1480 and 1497, and it gave a unique insight into what these individuals knew about each other. It removed much of the speculation about the existence of maps drawn by Cabot, where he landed in June 1497, and which part of the coastline he surveyed. It also stated that he gave names to many geographical features. The letter confirmed that Bristol's merchants had played a much greater role in the discovery of North America than had previously been recognized.

Day's letter stated that Bristol fishermen were known to be obtaining codfish in Newfoundland in the 1480s and further reveals that Columbus was fully informed in this regard. Day even questioned Columbus's ethics in claiming the New World for Spain, knowing that the Bristol merchants had been there before him.

Day's letter gives a remarkable background to Cabot's 1497 voyage to Newfoundland. It became the fifth account of the trip and shed new light on where he landed in North America. His description of "trees tall enough to make ships' masts" is strong evidence that he landed in Maine or Nova Scotia, not Newfoundland, as had been earlier supposed.

Day discusses the evidence of human habitation that they found:

> "They found a trail that went inland, they saw a site where a fire had been made, and they saw manure of animals which they thought to be farm animals, and they saw a stick half a yard long pierced at both ends, carved and painted with brazil, and by such signs they believe the land to be inhabited. Following the shore they saw two forms running on land one after the other; but they could not tell if they were human beings or animals; and it seemed to them that there were fields where they thought might also be villages…"

It documented that Cabot had made a map that covered 900 miles of coastline. It identified the Island of Seven Cities, the "mainland" of Brassyle, and names for all the major capes and other visible landmarks. Other names were added later after the return to Bristol, and the King himself renamed the Island of Seven Cities to "New founde land."

109

The most revealing piece of information contained in this letter is that the writer of the letter, Johan Day, sent a copy of Cabot's map to Christopher Columbus. This map, showing 900 miles of coastline complete with names and coordinates of the major capes and channels, was in Columbus's hands within a few months of Cabot's voyage.

It is almost certain that Columbus had Cabot's map before he left on his third voyage the following May and that Juan de la Cosa may have seen this, the 1497 map, and then later the 1498 map before he produced his chart in 1500 (showing five English flags and noting the areas discovered by the English). Vespucci, who was sailing in the same fleet as his friend de la Cosa in 1499 also would have had access to it.

☩

Ten years later in the 1960s, more documents came to light in the archives of The Mercer Company in Bridgnorth, Shropshire in England. They consisted of cargo manifests from several ships that sailed out of Bristol in the 1470s and 1480s. These seemingly mundane accounting records provided historians with a detailed record of the business dealings of many of the prominent merchants in Bristol:

> "Anno M iiii lxxx atte Mened the xxiiii day of Octobre Rychard Parkere Master of the Trenete
> Item. That y John Balsall delivered …to the Master x Rs a bord the schep at the Crosse montith
> Summa iii x M
>
> 1480 at Minehead (Somerset, England) 29 October, Richard Parker, Master of the Trinity
> I John Balsall delivered to the Master 10 Reales (Spanish silver coins) aboard the ship at The Cross (in Bristol)
> Amount paid out 310 Maravedis (Spanish coins)"

Amerike's name was one of many in the cargo manifests on the *Trinity* voyages to Lisbon in the late 1470s. He was also identified as the shipper of two tuns of salt and corrupt wine to Ireland in 1480 on one of Thomas Croft's ships. Many other prominent Bristolians were listed in these records, and enough information was extracted to build a rich history of life in Bristol during those years.

It became clear that Amerike not only was closely linked to Cabot's voyage in 1497 but also played a key role in voyages to Brassyle some seventeen years earlier. His involvement with America predated Columbus's discovery of the continent by twelve years.

This obscure figure, who was already known to have been Sheriff in 1501, now took on a new significance. His name also surfaced in the property deeds of three important land holdings: the Clifton Manor, Canynges Mansion on Redcliffe Street, and the Ashton-Phillips Estate.

Local historical documents suddenly took on a much greater significance. The annual civic records of the Bristol government had survived from the 15th and 16th centuries and were readily available for research to be carried out. There are two series of "Kalendars" in which local events were recorded by the town clerk of the City of Bristol. Ricart's Kalendar is the better known of the two. There are several examples from both series still in existence though neither is complete. A summary of these Kalendars was made in 1565 by Maurice Toby, and in this compilation he uses the name "America":

> "The land of America was found by the merchants of Bristow."

Under the mayoral year of 1496-1497, Toby recorded that John Drews was Mayor, Hugh Johnes was Sheriff, Thomas Vaughan and John Elyott were Bailiffs, and that:

> "This year (1497) on St John the Baptist's Day (June 24th), the land of America was found by the merchants of Bristow,

*in a ship of Bristowe called the "Mathew," the which said ship
departed from the port of Bristowe the 2nd of May and came
home again the 6th August following."*

His use of the term America reveals a familiarity with the word, and
in all likelihood it was written in the documents he had assembled and
from which he was quoting. Those referenced documents could have been
written much earlier, and in them the name America might actually have
had a different meaning. It could have been the name of a small settlement,
an island, or the unknown landmass the merchants saw in the distance. It
was probably not a reference to the whole continent. This Kalendar record
provides the most accurate dates for Cabot's successful voyage.

Unfortunately, the Toby manuscript, which was known and quoted
from in the nineteenth century, is no longer in existence. It was owned
in the 1800s by the Fust family of Hill Court, Gloucestershire, and sold.
It was eventually purchased by an antiquities dealer, Thomas Kerslake
of Park Street, Bristol. It was destroyed along with many other valuable
manuscripts when his shop caught fire in 1860.

17.

The Stars and Stripes

"Paly of six, Or and Azure, on a fess Gules, three mullets Argent."

RICHARD AMERIKE'S COAT OF ARMS,
THE STARS AND STRIPES

Today Bristol is a dynamic city of 400,000 people, but it is still possible to explore a medieval world that Richard Amerike would recognize. Hoefnagle's map, which shows Bristol as it was in 1580, can be used for a tour of the city today.

Bristol Bridge was at the center of Amerike's world. The structure in his day has long since disappeared, but the latest incarnation crosses the River Avon at the same location and leads straight into the same narrow city streets. The plan of the city has hardly changed, and within a quarter of a mile are the banking area, the law courts, the old stock exchange building, and the markets. Amerike would still recognize the quays with their moorings, some sixteen churches, and the market area with its labyrinth of alleyways, cellar bars, and eating houses.

To the northeast of the bridge is an entrance to the Castle area. Oliver Cromwell ordered the Castle to be disassembled 170 years after Amerike's death. Amerike's warehouses would have been on the quays of the River Avon, downstream from Bristol Bridge. Welsh Back, where the ships from South Wales docked, is still in use.

The house belonging to Arthur Kemys, which Cabot rented for his wife, stood on St. Nicholas Street, which winds around behind the market from St. Nicholas Church to Broad Quay. The shell of St. Nicholas Church, where both Cabot and Columbus worshipped, stands near Bristol Bridge. A German incendiary bomb destroyed it in the blitz on November 24, 1940. The church has been partially rebuilt and serves as a civic information center. Amerike's office in the Customs House looked out over the Broad Quay at the far end of St. Nicholas Street.

On King Street, the pub and eating house called the "Llandoger Trow" has been in operation for 400 years. The buildings in 1500 would have looked very similar. Draught sherry is served from wooden casks in this and several other historical pubs.

The sights on the half-mile walk from Bristol Bridge along Redcliffe Street to the magnificent St. Mary Redcliffe Church are not very attractive. It was an industrial area in Amerike's day, and it is no different today. At about the halfway point and fronting the river is the site of Canynges' mansion, which Amerike acquired and where his daughter Johanna lived. Some of the walls and a window from the original building are incorporated in the new office development there.

St. Mary Redcliffe Church can be seen from this vantage point. The spire is the only part of this building that does not date from the thirteenth century and Canynges' reconstruction. It was built in 1859 to replace the one that crashed down onto the nave in 1445. Today a visitor would have to cross the busy road to walk up the path to the entrance.

✠

PORT OF BRISTOL AND ST. MARY REDCLIFFE CHURCH

It is quiet inside the old church. The whalebone that John Cabot and the Bristol sailors dragged off a beach in Brassyle 500 years ago is in the small chapel to your right. You are now in the merchant traders' church and are in the presence of some of the people you have met in this book. Many of their weddings, christenings, religious celebrations, and funerals occurred within these walls.

St. Mary Redcliffe Church was over 200 years old when Richard Amerike attended the funerals of William Canynges and John and Joan Jay. Happier occasions would have been the wedding of Johanna to John Broke and the christenings of his grandchildren.

A carpet lies at the front of the choir just before the step up to the altar. Under the stone floor beneath this carpet, Johanna lies next to her husband, John Broke. The grave is marked by a brass plate that identifies her as Richard Amerike's daughter:

> "Here lies the body of that venerable man John Brook, sergeant at law of that most illustrious prince of happy memory King Henry the Eighth and Justice of Assize for the same King

115

in the Western parts of England, and Chief Steward of the honourable house and monastery of the Blessed Mary of Glastonbury in the County of Somerset which John died on the 25th day of December in the year of Our Lord one thousand five hundred and 22. And near him rests Johanna his wife one of the daughters and heirs of Richard Amerike on whose souls may God have mercy. Amen."

Alongside is the grave of John Jay, one of the owners of the *Trinity* that sailed to Lisbon so many years ago. He lies here with his wife, Joan, whose brother, William Worcestre, wrote about his brother-in-law's attempts to find the Island of Brassyle. Also buried with them are their fourteen children, including John Jay, who looked unsuccessfully for Brassyle in 1480. Their six sons and eight daughters are shown on the plaque. The depiction of the daughters with their hair pinned up indicates they died unmarried.

William Canynges, the man who controlled the codfish trade with Iceland, has his own large memorial.

In an ironic twist of fate, the whereabouts of Richard Amerike's gravesite is unknown. Given his standing in the community and his obvious associations with this church, it is hard to imagine that he is not buried here somewhere.

With no spelling conventions in his own lifetime, Richard Amerike's name was written many different ways. Sometimes it was ap Meric or ap Merryk. It appeared as Amerike in a land conveyance and on his daughter's grave. It showed up as Ameryk and Ap Meryke in the Westminster Abbey Customs Rolls. It was both Ameryke and Amyreke in the Balsall shipping accounts. The earlier spelling of ap Meric was reputedly the basis for the trademark that was used to identify his cargo.

Richard Amerike married Lucy Wells around the year 1460, and they had two daughters, Johanna being the eldest. Nothing is known of Richard Amerike's other daughter. The only reference to her is the engraving on Johanna's grave, indicating there was another daughter and that she was an heir to his estate.

A brass plaque on Johanna's grave was removed nearly 200 years ago, and the indent in the stone can be seen.[1] The description of the crest on the missing brass is that of the Amerike coat of arms:[2]

> "...in the chancel of St. Mary Redcliffe (Broke's) brass remains, also until recently (probably early 1800s), the arms of Cobham and Brook, quarterly, with a crescent for difference, as well as Brook impaling Americk, viz - Paly of six, or, and az. on a fess gu. three mullets arg."

In approximately 1486, Johanna married John Broke,[3] who was a member of the powerful Somerset family of Broke, in Ilchester. Somerset is the county immediately to the south of Bristol. Broke's grandfather, Thomas Broke, had been Lord Cobham. The title passed to the brother of John's father, who was the current Lord Cobham when John was born. Lord Cobham, Edmund (or Edward) Broke, was instrumental in King Edward IV's accession to the throne at the battle of Mortimer's Cross on February 3, 1461, the same battle that was fought on the land belonging to the family of Thomas Croft, Amerike's predecessor at the Customs House.

"You, Edward, shall to Edmund Brook Lord Cobham,
With whom the Kentishmen will willingly rise:
In them I trust, for they are soldiers,
Witty, courteous, liberal, full of spirit."

RICHARD DUKE OF YORK AND GOOD KING HENRY VI
THE BATTLE OF MORTIMER'S CROSS
WILLIAM SHAKESPEARE

Broke's eldest son, Thomas, was born in 1487. At that time, Amerike had just obtained the Canynges mansion in settlement of a lawsuit. He gave the mansion to Johanna and John Broke, who lived there for many years. John Broke was a lawyer and Sergeant-at-Law to the Royal Court in London. Amerike's house on his Ashton-Phillips Estate still stands, in Yanley Lane, in the nearby village of Long Ashton.

Half a mile away fron St. Mary Redcliffe Church and across the city center is Bristol Cathedral, which Amerike would remember as St. Augustine's Abbey. The Cathedral fronts College Green, one of Bristol's many open spaces. Also adjacent to College Green is Bristol's city government building, known as the Council House. John Cabot's statue looks toward the city from the center of the façade. A small church opposite the Cathedral is used exclusively by the Mayor and the city. It is called the Lord Mayor's Chapel. The Poyntz family tombs are in this building and Amerike's coat of arms, the "Stars and Stripes," is incorporated in three crests in this building.

Cabot Tower, built for the 1897 celebrations, is at the top of the hill behind the Council House. The Clifton Manor house, where Amerike lived for many years, is about a quarter of a mile to the left of the tower.

✠

Most visitors to Bristol[4] make a point of seeing the famous Clifton Suspension Bridge, which crosses the River Avon gorge, linking Clifton with the Ashton Estate. The bridge was constructed between 1830 and 1866 by the renowned nineteenth-century engineer, Isambard K. Brunel. Postcards picture the River Avon at high tide when it is undoubtedly more attractive. You should ignore Brunel's bridge for a moment and visit the river at low tide. You will then understand the significance of Columbus's recollections about Bristol's extreme tidal range.

However, the river level within the old city is no longer tidal. Two hundred years ago, the city built a series of locks and a new channel for the river so that the water in the city harbor can be kept permanently at high tide. The river now bypasses the original harbor in "The Cut." You will have to imagine how the river appeared at low tide in the center of the city.

About half a mile from the suspension bridge is the headquarters of the "Society of Merchant Venturers." This organization obtained its charter in 1552 and grew out of the Fellowship of Merchants. Today it primarily administers merchant marine charities.

The *Matthew* still sails the waters in Bristol Harbor. A replica of the ship that transported Cabot to Brassyle was built at the end of the twentieth century and was sailed from Bristol to Newfoundland in 1997, recreating Cabot's voyage exactly 500 years later. It is usually moored about a mile downstream from St. Mary Redcliffe Church and is next to the first of the modern propeller-driven, iron hulled ocean liners, the *S.S. Great Britain*. Both ships were built in Bristol's shipyards. The *Great Britain* was launched in 1842 and was also engineered by Brunel. At the time of this writing, it is undergoing restoration in the same dry dock where it was originally built.

Today, when the *Matthew* is not sailing between the coastal towns of southwest England, it carries visitors on scenic tours on the River Avon within the city of Bristol. Tourists can take this short voyage and imagine themselves as one of the eighteen people in the crew. The *Matthew* sails downriver from the city center, and heads to the west. It is easy to imagine Richard Amerike on board checking the cargo. There are fifty barrels of salt, thirty barrels of corrupt wine, leather hides, small pigs, several chickens, and enough food, beer, and firewood for seven months below deck in the hold. The fire is alight in the galley, where the meals are cooked.

The *Matthew* sails with the falling tide beneath the bluffs of Clifton, to the mouth of the Avon, and out into the current. The full sails are raised and fresh wind takes it out into the Bristol Channel and to the rough seas of the Mare Oceanum toward a distant small fishing village on the Island of Brassyle, which is now called America.

Epilogue

*The following clipping from the
Bristol and West Country daily newspaper,
"Western Daily Press," dated August 7, 1929,
was printed after Mr. Hudd's death.*

Recent references to the origin of the name given to the New World discovered by John Cabot in the Bristol ship *Matthew,* has aroused such widespread interest that we publish in full Mr. A. E. Hudd's paper on the subject.

The paper was read at a meeting of the Clifton Antiquarian Club on the 21st May, 1908, and clearly makes out a good case for further researches.

As already stated, the American Consul, Mr. Digby A. Willson, is greatly interested, and has asked for all the evidence available to submit to the appropriate authorities in the United States.

This and many other indications of interest in the subject, suggested that the publication of Mr. Hudd's statement in full, will encourage exhaustive inquiry into a matter of considerable importance to Bristol.

The full text of Mr. Alfred E. Hudd's address to the members of the Clifton Antiquarian Club is herewith given.

"...There is no longer any doubt on the return of [Cabot's] second voyage John received for the second time the handsome [sic] pension conferred upon him by the King, from the hands of the Collectors of Customs of the Port of Bristol. One of these officials, the senior of the two, who probably was the person who actually handed over the money to the explorer, was named Richard Ameryk (also written Ap Meryke in one deed) who seems to have been a leading citizen of Bristol at the time, and was Sheriff in 1503. Now it has been suggested both by Mr. Scott and myself that the name given to the newly-found land by the discoverer was "Amerika," in honour of the official from whom he received his pension."

Evidently, the United States government was contacted in 1929 regarding the Amerike theory. Perhaps the 1929 stock market crash, the Depression, and ultimately World War II relegated this item to the back burner.

Appendices

APPENDIX A

Dispatch of Pedros de Ayala
The Spanish Envoy in London
Dated 25 July 1498

I think your Highnesses have already heard how the King of England (Henry VII) has equipped a fleet to explore certain islands or mainland which he has been assured certain persons who set out last year from Bristol in search of the same have discovered. I have seen the map made by the discoverer (Cabot), who is another Genoese like Columbus, who has been in Seville and at Lisbon seeking to obtain persons to aid him in this discovery. For the last seven years the people of Bristol have equipped two, three and four caravels to go in search of the islands of Brazil and Seven Cities according to the fancy of this Genoese. The king made up his mind to send thither, because last year sure proof was brought him they had found land. The fleet he prepared, which consisted of five vessels, was provisioned for a year. News has come that one of these, in which sailed another, Friar Buil (who sailed with Columbus in his second voyage), has

125

made land in Ireland in a great storm with the ship badly damaged. The Genoese kept on his way. Having seen the course they are steering and the length of the voyage, I found that what they have discovered or are in search of is possessed by Your Highnesses because it is at the cape which fell to Your Highnesses by the convention with Portugal (Treaty of Tordesillas, 1494). It is hoped they will be back by September. I will let Your Highnesses know about it. The king has spoken to me several times on the subject. He hopes the affair may turn out profitable. I believe the distance is not 400 leagues. I told him I believed the islands were those found by Your Highnesses, and although I gave him the main reason, he would not have it. Since I believe Your Highnesses will already have notice of all this and also of the chart or mappemonde which this man has made, I do not send it now, although it is here, and so far as I can see exceedingly false, in order to make believe that these are not part of the said islands (of Your Highnesses).

REPRODUCED FROM H. P. BIGGAR

APPENDIX B

TRANSLATION OF THE JOHAN DAY LETTER TO COLUMBUS, THE LORD GRAND ADMIRAL

Your Lordship's servant brought me your letter. I have seen its contents and would be most desirous and most happy to serve you. I do not find the book Inventio Fortunada and I thought that he was bringing it with my things, and I am very sorry not to find it because I wanted very much to serve you. I am sending another book of Marco Polo and a copy (map) of the land which has been found. I do not send the map because I am not satisfied with it for the many occupations forced me to make it in a hurry at the time of my departure; but from the said copy your Lordship will learn what you wish to know: for in it are named the capes of the mainland and the islands, and thus you will see where the land was first sighted, since most of the land was discovered after turning back.

Thus your Lordship will know that the cape nearest to Ireland is 1800 miles west of Dursey Head which is in Ireland, and the southernmost part of the Island of Seven Cities (Newfoundland?) is west of Bordeaux River, and your Lordship will know that he landed at only one spot of the mainland, near the place where land was first sighted; and they disembarked there with a crucifix and raised banners and the arms of the Holy Father; and those of the King of England, my master, and they found tall trees of the kind masts are made, and other smaller trees, and the country is very rich in grass.

In that particular spot, as I told your Lordship, they found a trail that went inland, they saw a site where a fire had been made, and they saw manure of animals which they thought to be farm animals, and they saw a stick half a yard long pierced at both ends, carved and painted with brazil, and by such signs they believe the land to be inhabited. Since he was with just a few people, he did not dare advance inland beyond the shooting distance of a cross-bow, and after taking in fresh water he returned to his ship.

All along the coast they found many fish like those which in Iceland are dried in the open and sold in England and other countries, and these fish are called in English "stockfish": and thus following the shore they saw two forms running on land one after the other; but they could not tell if they were human beings or animals; and it seemed to them that there were fields where they thought might also be villages, and they saw a forest whose foliage looked beautiful.

They left England toward the end of May, and must have been on the way 35 days before sighting land; the wind was east-north-east and the sea calm going and coming back, except for one day when he ran into a storm two or three days before finding land; and going so far out his compass needle failed to point north and marked two rhumbs below. They spent about one month discovering the coast and from the above mentioned cape of the mainland which is nearest to Ireland, they returned to the coast of

Europe in fifteen days. They had the wind behind them, and he reached Brittany because the sailors confused him, saying that he was heading too far north. From there he came to Bristol, and he went to see the King to report to him all of the above mentioned; and the King granted him a pension of twenty pounds sterling to sustain himself until the time comes when more will be known of this business, since with God's help it is hoped to push through plans for exploring the said land more thoroughly next year with ten or twelve vessels - because in his voyage he had only one ship of 50 'toneles' and twenty men and food for seven or eight months - and they want to carry out this new project.

It is considered certain that the cape of the said land was found and discovered in the past by the men from Bristol who found "Brazil" as your Lordship well knows. It was called the Island of Brazil, and it is assumed and believed to be the mainland that the men from Bristol found.

Since your Lordship wants information relating to the first voyage, here is what happened: he went with one ship, his crew confused him, he was short of supplies and he ran into bad weather, and he decided to turn back.

Magnificent Lord, as to other things pertaining to the case, I would like to serve your Lordship if I were not prevented in doing so by the occupations of great importance relating to shipments and deeds for England which must be attended to at once and keep me from serving you, but rest assured Magnificent Lord, of my desire and natural intention to serve you, and when I find myself in other circumstances and more at leisure, I will take pains to do so; and when I get news from England about the matters referred to above - for I am sure that everything has to come to my knowledge - I will inform your Lordship of all that would not be prejudicial to the King my master. In payment of some services which I hope to render you, I beg your Lordship to kindly write me about such matters, because the favour you will thus do me will greatly stimulate my memory to serve you in all the things that may come to my knowledge. May God keep prospering your Lordship's magnificent state according to your merits. Whenever your Lordship should find it convenient, please remit the book or order it to be given to Master George.

I kiss your Lordship's hands
Johan Day

TRANSLATED BY DR. LOUIS VIGNERAS, 1956

APPENDIX C

LETTERS OF JOHN CABOT'S VOYAGES

LETTER FROM LORENZO PASQUALIGO TO HIS BROTHERS ALVISE AND FRANCESCO. LONDON, 23RD AUGUST, 1497

Our Venetian, who went with a small ship from Bristol to find new islands, has come back, and says he has discovered, 700 leagues off, the mainland of the country of the Gran Cam, and that he coasted along it for 300 leagues, and landed, but did not see any person. But he has brought here to the king certain snares spread to take game, and a needle for making nets, and he found some notched trees, from which he judged that there were inhabitants. Being in doubt, he came back to the ship. He has been away three months on the voyage, which is certain, and, in returning, he saw two islands to the right, but he did not wish to land, lest he should lose time for he was in want of provisions. This king has been much pleased. He says that the tides are slack, and do not make currents as they do here. The king has promised for another time, ten armed ships as he desires, and has given him all the prisoners, except such as are confined for high treason, to go with him, as he has requested; and has granted him money to amuse himself till then. Meanwhile, he is with his Venetian wife and his sons at Bristol. His name is Zuam Talbot, and he is called the Great Admiral, great honor being paid to him, and he goes dressed in silk. The English are ready to go with him, and so are many of our rascals. The discoverer of these things has planted a large cross in the ground with a banner of England, and one of St. Mark, as he is a Venetian; so that our flag has been hoisted very far away.

✠

Extract from the First Dispatch of Raimondo di Soncino to the Duke of Milan. 24th August, 1497

Some month afterwards His Majesty sent a Venetian, who is a distinguished sailor, and who was much skilled in the discovery of new islands, and he has returned safe, and has discovered two very large and fertile islands, having, it would seem, discovered the seven cities 400 leagues from England to the westward. These successes led His Majesty at once to entertain the intention of sending him with fifteen or twenty vessels.

✠

Second Dispatch of Raimondo di Soncino to the Duke of Milan. 18th December, 1497

My most illustrious and most excellent Lord,

Perhaps amidst so many occupations of your Excellency it will not be unwelcome to learn how this Majesty has acquired a part of Asia without drawing his sword. In this kingdom there is a certain Venetian named Zoanne Caboto, of gentle disposition, very expert in navigation, who, seeing that the most serene Kings of Portugal and Spain had occupied unknown islands, meditated the achievement of a similar acquisition for the said Majesty. Having obtained royal privileges securing to himself the use of the dominions he might discover, the sovereignty being reserved to the Crown, he entrusted his fortune to a small vessel with a crew of 18 persons, and set out from Bristo, a port in the western part of this kingdom. Having passed Ibernia, which is still further to the west, and then shaped a northerly course, he began to navigate to the eastern part, leaving the North Star on the right hand; and having wandered thus for a long time, at length he hit upon land, where he hoisted the royal standard, and took possession for his Highness, and, having obtained various proofs of his discovery, he returned. The said Messer Zoanne, being a foreigner and poor, would not have been believed if the crew, who are nearly all English, and belonging to Bristo, had not testified that what he said was the truth. This Messer Zoanne has the description of the world on a chart, and also on a solid sphere which he has constructed, and on which he shows where he has been; and, proceeding towards the east, he has passed as far as the country of the Tanais. And they say that there the land is excellent and (the climate?) temperate, suggesting that brasil and silk grow there. They affirm that the sea is full of fish, which are not only taken with a net, but also with a basket, a stone being fastened to it in order to keep it in the water; and this I have heard stated by the said Messer Zoanne.

The said Englishmen, his companions, say that they took so many fish that this kingdom will no longer have need of Iceland, from which country there is an immense trade in the fish they call stock-fish. But Messer Zoanne has set his mind on higher things, for he thinks that, when that place has been occupied, he will keep on still further towards the east, where he will be opposite to an island called Cipango, situated in the equinoctial region, where he believes that all the spices of the world, as well as the jewels, are found. He further says that he was once at Mecca, whither the spices are brought by caravans from distant countries; and having inquired from whence they were brought and where they grow, they answered that they did not know, but that such merchandize was brought from distant countries by other caravans to their home; and they further say that they are also conveyed from other remote regions. And he adduced

131

this argument, that if the eastern people tell those in the south that these things come from a far distance from them, presupposing the rotundity of the earth, it must be that the last turn would be by the north towards the west; and it is said that in this way the route would not cost more than it costs now, and I also believe it. And what is more, this Majesty, who is wise and not prodigal, reposes such trust in him because of what he has already achieved, that he gives him a good maintenance, as Messer Zoanne has himself told me. And it is said that before long his Majesty will arm some ships for him, and will give him all the malefactors to go to that country and form a colony, so that they hope to establish a greater depot of spices in London than there is in Alexandria. The principal people in the enterprise belong to Bristo. They are great seamen, and, now that they know where to go, they say that the voyage thither will not occupy more than 15 days after leaving Ibernia. I have also spoken with a Burgundian, who was a companion of Messer Zoanne, who affirms all this, and who wishes to return because the Admiral (for so Messer Zoanne is entitled) has given him an island, and has given another to his barber of Castione, who is a Genoese, and both look upon themselves as Counts; nor do they look upon my Lord the Admiral as less than a Prince. I also believe that some poor Italian friars are going on this voyage, who have all had bishopricks promised to them. And if I had made friends with the Admiral when he was about to sail, I should have got an archbishoprick at least; but I have thought that the benefits reserved for me by your Excellency will be more secure. I would venture to pray that, in the event of a vacancy taking place in my absence, I may be put in possession, and that I may not be superseded by those who, being present, can be more diligent than I, who am reduced in this country to eating at each meal ten or twelve kinds of victuals, and to being three hours at table every day, two for love of your Excellency, to whom I humbly recommend myself.

London, 18 Dec. 1497, your Excellency's most humble servant, Raimundus.

The text of these letters is from the Hakluyt Society's edition of Columbus's Journal. This text is provided by the Internet Modern History Sourcebook.

APPENDIX D

THE FOLLOWING EXCERPTS ARE TAKEN FROM A LETTER FROM AMERIGO VESPUCCI TO HIS CHILDHOOD FRIEND PIERO SODERINI, GONFALONIER OF THE REPUBLIC OF FLORENCE.
AN ACCOUNT OF THE FIRST OF HIS FOUR VOYAGES.

Magnificent Lord. After humble reverence and due commendations, etc. It may be that your Magnificence will be surprised by my rashness and your customary wisdom, in that I should so absurdly bestir myself to write to your Magnificence the present so-prolix letter: knowing that your Magnificence is continually employed in high councils and affairs concerning the good government of this sublime Republic...

...that the King Don Ferrando of Castile being about to dispatch four ships to discover new lands towards the west, I was chosen by his Highness to go in that fleet to aid in making discovery: and we set out from the port of Cadiz on the 10th day of May 1497, and took our route through the great gulf of the Ocean-sea: in which voyage we were eighteen months: and discovered much continental land and innumerable islands, and great part of them inhabited: whereas there is no mention made by the ancient writers of them: I believe, because they had no knowledge thereof...

...westward, according to the shewing of our instruments, 75 degrees from the isles of Canary: whereat we anchored with our ships a league and a half from land; and we put out our boats freighted with men and arms: we made towards the land, and before we reached it, had sight of a great number of people who were going along the shore: by which we were much rejoiced: and we observed that they were a naked race...

Amongst those people we did not learn that they had any law, nor can they be called Moors nor Jews, and (they are) worse than pagans: because we did not observe that they offered any sacrifice: nor even had they a house of prayer: their manner of living I judge to be Epicurean: their dwellings are in common: and their houses (are) made in the style of huts, but strongly made, and constructed with very large trees, and covered over with palm-leaves, secure against storms and winds: and in some places (they are) of so great breadth and length, that in one single house we found there were 600 souls: and we saw a village of only thirteen houses where there were four thousand souls: every eight or ten years they change their habitations: and when asked why they did so: (they said it was) because of the soil which, from its filthiness, was already unhealthy and corrupted, and that it bred aches in their bodies, which seemed to us a good reason: their riches consist of bird's plumes of many colors, or of rosaries which they make from fishbones, or of white or green stones which they put in their cheeks and in their lips and ears, and of many other things which we in no wise value: they use no trade, they neither buy nor sell. In fine, they live and are contended with that which nature gives them. The wealth that we enjoy in this our Europe and elsewhere, such as gold, jewels, pearls, and other riches, they hold as nothing; and although they have them in their own lands, they do not labor to obtain them, nor do they value them. They are liberal in giving, for it is rarely they deny you anything: and on the other hand, liberal in asking, when they shew themselves your friends...

...the men of the canoes drew away from us, and began with their bows to shoot arrows at us: and those who were swimming each carried a lance held, as covertly as they could, beneath the water: so that, recognizing the treachery, we engaged with them, not merely to defend ourselves, but to attack them vigorously, and we overturned with our boats many of their almadie or canoes, for so they call them, we made a slaughter of them, and they all flung themselves into the water to swim, leaving their canoes abandoned, with considerable loss on their side, they went swimming away to the shore: there died of them about 15 or 20, and many were left wounded: and of ours 5 were wounded, and all, by the grace of God, escaped death: we captured two of the girls and two men: and we proceeded to their houses, and entered therein, and in them all we found nothing else than two old women and a sick man: we took away from them many things, but of small value: and we would not burn their houses, because it seemed to us a burden upon our conscience: and we returned to our boats with five prisoners: and betook ourselves to the ships, and put a pair of irons on the feet of each of the captives, except the little girls...

Your Magnificence must know that in each of my voyages I have noted the most wonderful things, and I have indited it all in a volume after the manner of a geography: and I entitle it Le Quattro Giornate: in which work the things are comprised in detail, and as yet there is no copy of it given out, as it is necessary for me to revise it. This land is very populous, and full of inhabitants, and of numberless rivers, animals: few resemble ours, excepting lions, panthers, stags, pigs, goats, and deer: and even these have some dissimilarities of form: they have no horses nor mules, nor, saving your reverence, asses nor dogs, nor any kind of sheep or oxen: but so numerous are the other animals which they have, and all are savage, and of none do they make use for their service, that they could not be counted. What shall we say of others such as birds? which are so numerous, and of so many kinds, and of such various-coloured plumages, that it is a marvel to behold them.

...and when the next day arrived, we beheld coming across the land a great number of people, with signals of battle, continually sounding horns, and various other instruments which they use in their wars: and all painted and feathered, so that it was a very strange sight to behold them: wherefore all the ships held council, and it was resolved that since this people desired hostility with us, we should proceed to encounter them and try by every means to make them friends: in case they would not have our friendship, that we should treat them as foes, and so many of them as we might be able to capture should all be our slaves: and having armed ourselves as best we could, we advanced towards the shore, and they sought not to hinder us from landing, I believe from fear of the cannons: and we jumped on land, 57 men in four squadrons, each one a captain and his company: and we came to blows with them: and after a long battle many of them were slain, we put them to flight, and pursued them to a village, having made about 250 of them captives, and we burnt the village, and returned to our ships with victory and 250 prisoners, leaving many of them dead and wounded, and of ours there were no more than one killed and 22 wounded, who all escaped, God be thanked. We arranged our departure, and seven men, of whom five were wounded, took an island-canoe, and with seven prisoners that we gave them, four women and three men, returned to their country full of gladness, wondering at our strength: and we thereon made sail for Spain with 222 captive slaves: and reached the port of Calis (Cadiz) on the 15th day of October, 1498, where we were well received and sold our slaves. Such is what befell me, most noteworthy, in this my first voyage.

TRANSLATION FROM VESPUCCI'S ITALIAN, PUBLISHED IN FLORENCE IN 1505-6 BY "M. K." FOR QUARITCH'S EDITION, LONDON, 1885

APPENDIX E

Bristol Evening Post, Friday April 30, 1943

AMERYK AND AMERICA

Although we like to think that Cabot discovered America- and there seems historical evidence in support of the fact- most Americans seem to favour Columbus' claim.

I was not a little surprised therefore, to hear Gen. H. R. Ingles, presenting the Stars and Stripes to the Lord Mayor, say: "Cabot was really the discoverer of what is now known as America."

Concerning that statement, I have to thank Mr. H. R. Simpson (M.A., Camb.) for an interesting contribution.

"I think you could assure Americans that they owe not only their name, but their flag to Bristol," he writes. "The accepted derivation of the word America from Amerigo Vespucci is based on evidence so slight as to be almost fantastic. The name America was given by Richard Ameryk, Lord of the Manor of Clifton, and sometime Sheriff and Receiver of the King's Customs at Bristol.

Ameryk, from his connection with Canynge (his sister seems to have been that worthy's mistress), undoubtedly shared the common Bristol interest in discovery: it is even possible that he held a patent [trading license] from Edward IV giving him a title to anything that could be found in the West.

FINANCED CABOT

In any event, he financed Cabot, possibly with the help of his son-in-law, Broke, and it seems inevitable that he should have stipulated that any new continent disclosed should bear his name.

Cabot, an unreliable character, broke the bargain. He called his discovery Newfoundland- not an inspired name. Ameryk took the only revenge open to him. When as Receiver of the King's Customs, he was ordered to pay Cabot the Royal reward of £10, Ameryk refused to pay. Finally, when a very summary order came from the Exchequer, he did pay- and it is shortly after that that the name America begins to be used.

Any encyclopaedia will give the derivation of the Stars and Stripes, but the fact remains that the arms of the Ameryk family were stars and stripes. A curious coincidence, to say the least of it.

Notes

Chapter 1

1. It was a serious crime to give away state secrets from the Map Depository in Portugal, a crime that was punishable by death. Vespucci was reprimanded on one occasion for breaching the rules.

Chapter 2

1. The spice trade involved the cooperation of people from different cultures as the goods were sold from merchant to merchant on their journey west. At the far eastern end, the Chinese collected cloves and nutmeg from the East Indies and delivered them to the Malaysian port of Malacca. Muslim merchants from India, Malacca, and Arabia transported them across the Bay of Bengal to India. There, the cinnamon from Ceylon and locally produced pepper were added to the cargo and sold in the spice ports of Calicut, Cochin, Cannore, Goa, and Gujerat along the western coast of India. From India, the goods were sent to Persia, Arabia, and East Africa. Most of the cargo was taken by boat through the Persian Gulf to Shatt-al-Arab. The goods were then taken by riverboat and camel

caravan to Aleppo, Damascus, and eventually Constantinople. Other overland caravans from Cathay through Tibet and Persia also converged on Constantinople.

Chapter 3

1.   Christopher Columbus's journals contain two references to his early visit to England, as well as a specific reference to the Irish port of Galway where he saw two bodies (possibly Eskimos) that had washed up on the beach:

> *"Men of Cathay, which is toward the Orient, have come hither. We have seen many remarkable things especially in Galway, in Ireland, a man and a woman of most unusual appearance have come to land in two boats."*

He referred to Iceland, which he called Tile or Thule:

> *"To this island, which is as big as England, the English merchants go, especially those from Bristol."*

He also described the extreme fifty foot tides that he could only have observed in Bristol and the nearby River Severn:

> *"At the time I was there the sea was not frozen, but there were vast tides, so great that they rose and fell as much as twenty-six fathoms in depth."*

His journals also include a detailed description of the distinctive butterfly shape of an island he saw on that trip near the Arctic Circle. That description matches the outline of San Jan Maya Island, 400 miles north of Iceland, where Bristol ships regularly fished for cod:

> *"In February 1477, I sailed myself a hundred leagues beyond Thule whose northern point is seventy-three degrees distant from the equinoctial."*

2.   The tidal variations in the Mediterranean Sea are minimal, never rising more than a few feet. They are scarcely any different in the Atlantic Ocean, Lisbon, or Iceland. However, the sea between England and South Wales, and the approach to Bristol, has an extremely high tidal variation that can be as great as fifty feet in March and September. This is a major problem for shipping, especially for the unwary. Bristol's location is unique in Europe in that it is situated at the narrow end of a long funnel formed by the Bristol Channel, the estuary of the River Severn, and the River Avon. As the tide turns, the waters surge into the funnel and flood the city's port, and the harbor is drained equally rapidly as the tide recedes.

At the time, Bristol would have been the only location where a phenomenon such as this could have been observed. The other places in the world that experience such extreme tidal variations are outside Europe. They include the Bay of Fundy in Canada, an estuary in China, and the Cook Inlet in Alaska, all of which have a similar geographical configuration.

3.   The Canynges Company traded almost exclusively in the northern markets in the Baltic Sea and Scandinavia, and was the major Bristol-based shipping business. It was also the largest business in the city, owning a fleet of ten merchant ships.

The *Mary Redcliffe* had a 500-tun capacity, and the *Mary and John*, which was sailing in the 1450s, was a huge 900-tun ship that had cost Canynges £2,660. He is thought to have purchased this particular ship in the Baltic. Another ship, the *Mary Canynges*, may have been named after his daughter. Headed by its founder, William Canynges, it employed 800 sailors and an additional 100 people in its shipyards.

In the 1450s and 1460s, a Canynges ship would regularly sail from Bristol to Iceland. Several merchants would consign cargo on board, and it would be traded almost entirely for salt cod and stockfish.

4. The many trademarks that have been cataloged in the last hundred years were mostly preserved in seals or recorded on documents used by the merchants.

CHAPTER 4

1. Exports from Bristol to Portugal, Spain, Gascony, and Ireland

FOR THE PERIOD SEPT. 29, 1479, TO JULY 3, 1480

DESTINATION	SAILINGS	BROADCLOTHS (48 SQ. YDS.)
PORTUGAL	4	441
SPAIN	8	950
GASCONY	7	315
IRELAND	31	492
ICELAND	2	

Ninety to ninety-five percent of the exports to Portugal, Spain, and Gascony were cloth.

PORT OF BRISTOL RECORDS, 1480

Chapter 5

1. Imports to Bristol from Portugal, Spain, and Gascony

For the period Sept 29, 1479, to July 3, 1480

	Portugal	Spain	Gascony
Wine	206 tuns	197 tuns	816 tuns
Woad	£4	£424	£2,448
Iron		£285	£94
Oil	£923	£169	
Sugar	£426	£100	
Wax	£209	£20	
Soap		£63	
Fruit	£191	£4	
Salt	£83		
Other goods	£87	£114	£31
Total (exc. wine)	£1,924	£1,179	£2,573

Chapter 6

1. Increase in imports from Spain between the years 1480 and 1493:

	Yr to Oct 1480	Yr to Oct 1493
Wine	270 tuns	746 tuns
Woad	£570	£3,042
Iron	£380	£1,641
Oil	£245	£1,282
Sugar	£135	£28

Wax	£27	£105
Soap	£84	
Fruit	£5	£46
Salt		
Other goods	£150	£351
Total	£1,596	£6,495
(exc. wine)		

2. The Trinity, Lisbon to Bristol, Arr. March 4, 1480

The Crew

Captain	
Purser	
Ordinary seamen	18
Soldiers	8
Gunners	2
Cabin boys	3

Merchants	Cargo
Richard Parker	Wine, 76 tuns
John Balsall	Oil, 182 tuns Salt, 2 tuns Sugar, 54 cwt
Robert Strange	Wine Wax, 60 cwt

JOHN PYNKE	OIL
	VINEGAR
	WINE
	SALT
	VINEGAR
	MISC
RICHARD AMERIKE	OIL
	SUGAR
WILLIAM BIRD	OIL
	WAX
JOHN ESTERFIELD	OIL
	WINE
WILLIAM WODYNGTON	OIL
	SUGAR
	WINE

PLUS 92 OTHERS

3.   More circumstantial evidence came to light in the 1960s when some detailed shipping records for the *Trinity* were accidentally discovered at the Mercer Company in Bridgnorth. Richard Amerike's name repeatedly appears in bills of lading in these records in his earlier role as a merchant trader. These records produced a long list of merchants who specialized in the Spanish and Portuguese markets. When it was compared with the merchants who traded in the northern markets, very few names were repeated.

4.   Richard Amerike was most likely born either in Bristol or the Welsh border country that was voluntarily within English control. His

Welsh father or perhaps his grandfather may have moved to Bristol or Monmouthshire and registered the surname "Ap Meric" with the authorities. If Richard Amerike had been born in West Wales, his father's name would have been Meric Ap (his father's given name). There would not have been a surname. Like the King, Amerike could also trace his ancestry back to the royal houses of Wales.

Richard Amerike was a direct descendant of Einion Yrth ap Cunedda, who was King of Gwynedd and Anglesey, in North Wales, in the fifth century. Einion was succeeded by Caswallon Llaw-Hir who ruled for many years until 517. This royal line continued through Maelgwyn Hir Gwynedd ap Caswallon, 517-549, Rhun Hir ap Maelgwyn Hir Gwynedd, 549-586, and Beli ap Rhun, 586-599.

This lineage can be traced back even further to Beli Mawr, the King of Britain in about 100 B.C. whose son, Caswallon, was the general who opposed Julius Caesar's troops when the Roman army invaded Britain.

In 1343, Elizabeth, also a descendent of Beli ap Rhun, married Sir John Poyntz, the wealthiest and most politically powerful person in the Bristol area at that time. Both Queen Elizabeth II and the late Diana, Princess of Wales, were descendants of the Poyntz family via their Tudor lineage and the Spencer family connections.

Both Howel ap Meredydd and Elizabeth Poyntz share common ancestry with Richard Amerike.

The Poyntz family tomb is located in the Lord Mayors Chapel, which is situated across College Green from Bristol Cathedral. There are three Poyntz family crests in the church and on tombs that incorporate the Amerike coat of arms, attesting to this marriage connection. The Amerike crest is one-fourth of one crest, one-sixth of another, and one-tenth of the third.

The Merrick family, well established in the area around Sherbourne in Dorset and Somerset in the 1500s were descended from the same family.

5. In the late 1400s, Wales was a country that had been invaded and occupied by the English. The English authorities were administering the

country and required everyone to register a surname, in English, to avoid the confusion that reigned. Many Welsh surnames date from that time, when the last name was anglicized and used thereafter by the family as its surname. Examples of this are Parry, which began as ap Harry, and Probert, which began as ap Robert.

Welshmen who moved to England or lived in the border regions, or marches, faced this situation earlier. Amerike's father or grandfather may have moved to Bristol in the early 1400s, and the name was established when he registered his name with the authorities. It was subsequently anglicized. Thus, Ap Meric became Amerike.

CHAPTER 7

1. The profits made by reselling dried salt cod in Spain and Portugal were much greater than those made by importing it to England. The mild climate and availability of fresh and salted fish from Ireland met most of the English market needs. However, the hot climate in Spain and Portugal and the doctrines of the Catholic Church meant that this market had virtually no fish supplies if dried cod was unavailable.

2. Basque Menu for Bacalao (Salt Cod)
Main dish
Yields: 4 servings

> 2 LB SALT COD
> 1/4 C OLIVE OIL
> 2 MEDIUM ONIONS
> I SMALL CAN TOMATO SAUCE
> 2 CLOVES GARLIC
> 1/2 CUP MINCED PARSLEY
> 4 OZ SLICED PIMENTOS
> 1/2 CUP DRY WHITE WINE

Wash the cod, changing the water three or four times. Cover in water and soak overnight. Drain. Cut the fish into 2-inch pieces. Place the pieces in cold water to cover, bring to boil, and drain well. Chop the onions and garlic and fry them, along with the drained pimentos, in olive oil. When the vegetables are tender, stir in the tomato sauce, parsley, and wine. Cover and simmer over very low heat for 30 minutes. Place the cod in the sauce and simmer, uncovered, 30 to 45 minutes, or until the fish flakes easily.

http://www.recipesource.com/ethnic/europe/basque/bacalao1.html

3. De la Founte's ship, the *Christopher*, often sailed from Bristol to Lisbon and back, and then on to Iceland with the same skipper. On one particular voyage, both William Spencer and John Pynke shipped fruit from Portugal to Bristol. De la Founte was well positioned to supply salt cod to the Spanish and Portuguese market from his own suppliers in Iceland.

The *Christopher* would later achieve notoriety, in 1485, when it was captured by Inishbofin pirates off the coast of Galway, triggering an English-Irish diplomatic crisis.

4. There is no record of Richard Amerike ever trading in the Icelandic market. He employed an Icelandic servant in his household as evidenced by a record of him paying a 2-shilling "foreign worker subsidy" in a census taken in 1485. There were forty-eight Icelandic servants in Bristol that year.

5. Trade with Iceland did not cease entirely after 1475, though the impact was dramatic on the economies of Iceland and Bristol. There is a record of two voyages from Bristol to Iceland in 1480.

6. This marked the end of the Canynges Company's domination of Bristol business. William Canynges had died in 1474 at the age of 75. He had no dependents, so his estate had been split up among his relatives.

A year after his death, one of the Canynges fleet, along with a relative, Thomas Canynges, fell into the hands of Breton pirates who demanded a ransom of £100. The family turned to Richard Amerike to borrow the ransom money.

There is a record of Amerike later suing the Canynges heirs for nonpayment of this debt. Amerike obtained the Canynges mansion on Redcliffe Street in settlement. The house was situated on the river near St. Mary Redcliffe Church and later became the home of Amerike's daughter Johanna and her husband, John Broke.

7.   Basque fishermen may have established settlements in Newfoundland in the late 1400s, and Bristol merchant fishermen were there by 1480. A Danish sailor named Dietrich Pining also claimed to have discovered Newfoundland in 1472. In 1532, the French explorer Jacques Cartier planted a cross on the Gaspe Peninsula in the St. Lawrence River in Quebec and claimed it for France. He noted the presence of 1,000 Basque fishing vessels.

CHAPTER 8

1.   It is probable that this *Trinity* was not the 360-tun vessel that Worcestre wrote of in 1480 but was a smaller fishing ship that fit the terms of the trading license, which limited the size of the ships to 60 tuns.

2.   Much has been made of the fact that Richard Amerike was not one of the jurors in Croft's trial or on the list of forty-four potential jurors. Before the trial started, he was selected to relate his knowledge of the case. He was later eliminated, suggesting he was an interested party. Strange, Spencer, and de la Founte, the other three license holders, were also eliminated from the beginning.

3.   When the trading license expired in June 1483, the trade route to Brassyle had been established.

4.   Since Amerike was obviously financing cargo to Brassyle on a regular basis and would have paid for the construction of any shelters on the beach, the map in Sturmy's House may have had his name written on this harbor. The rutters prepared for skippers hauling Amerike's salt and goods over many years may have repeatedly used this name as the unloading point for the cargo.

5.   Barrels of salt and casks of wine each marked with Amerike's trademark may have been shipped to Brassyle as early as 1481. These same barrels would return to Bristol full of salted fish.

6.   The rutters for a voyage to Brassyle would have been elaborate. The prevailing winds at the latitude of England and Scotland are from the southwest. The rutters would dictate a voyage north to Iceland, following coastal landmarks around southern Greenland, and across the channel to the Labrador coast. The returning ship could return to Bristol with the prevailing winds and currents across the open ocean, confident of reaching land in about fifteen days.

7.   Merchant traders from Bristol had found the island of Brassyle and the world's richest fishing grounds, situated 1,200 miles out in the Mare Oceanum. When Christopher Columbus heard about these voyages a few years later, it would be obvious to him that the Bristol fishing ships had, in fact, reached "the northeast coast of Cathay."

Chapter 9

1.   The master mariners of Amerike's day relied on a great deal of personal skill in finding their destinations. They had very few tools to work with.

They needed to be able to measure five things accurately to determine where they were in relation to the land. These were north, speed of the ship, time, ship's latitude, and depth of water. They were unable to measure longitude except by estimating distances traveled and the maritime skills and intuition they developed.

NORTH: The magnetic compass was usually a crude affair and might be a piece of lode-stone resting on a piece of wood or a magnetized needle inside a straw, which was floated in a bucket of water. This compass was fairly accurate in Europe because magnetic north lines up approximately with true north. This however is not true in other parts of the world. From North America, it gives a very distorted "north" because the pole of magnetic north is far out of line with true north. This is why early maps show the Eastern Seaboard of North America running almost east to west, rather than southwest to northeast.

SPEED: The speed of the vessel was measured by tossing a piece of wood overboard at the bow and timing how long it took to get to the stern. This method was later improved upon by adding a long length of rope with knots tied at specific points. The speed was calculated by the number of knots that passed through one's hands in a given time.

TIME: An hour glass was essential to measure time. Time was continuously measured, and the hour was logged on a Peg-Board. The time was compared with the noon readings of the sun and the expected time of sunrise and sunset.

LATITUDE: The quadrant was used to measure the line of latitude on which they were positioned. It consisted of a heavy metal plate with an attached protractor that was hung in the rigging, with a plumb line hanging down to the deck. The navigator sighted along the protractor's edge at either the sun or the North Star and read the angle.

WATER DEPTH: A plumb line with a heavy lead weight was hung over the bow to constantly monitor the depth of the water when they were near the coast. The weight had an indent that allowed the crew to sample the sea bed.

A captain about to take a vessel on an established trade route was given the rutters for that voyage. This handwritten book contained a detailed description of the trip, including every landmark, heading, distance, and any and every detail about tides and prevailing winds that was available. It gave specific directions from landmark to landmark, and it would lead him to his destination. Before the age of printing, no charts were available other than those produced by the mariners themselves.

The rutters, the cargo manifest, and any other travel documents needed were given to the captain before he left port.

2.   Amerike's annual salary was £6-13s-4d, a relatively small sum that was the equivalent of a skilled craftsman's wage. The position was an honorary appointment, and he was responsible for the collection of import taxes and Crown payments from revenues collected in Bristol. The Customs House, where he and his two bailiffs had their offices, was situated on the Broad Quay, where the larger ships tied up.

3.   Robert, the eldest son of Robert Thorne, wrote that his father and Hugh Elliot visited Newfoundland several years before the Cabot voyage (believed to be in 1494). The word "discover" did not have the same meaning in 1500 as it does today. "Visit" would be the equivalent term. Also, Newfoundland is used in a more general context. Thorne writes with some regret that they were in a position to sail further west and south along the American coast, but the crew didn't want to cooperate.

Robert Thorne, who sailed with Cabot in 1497, would write to a Dr. Leigh in 1514 regarding his involvement:

> *"this inclination and desire of this discovery I inherited from*
> *my Father, who with another merchant of Bristol, named*

Hugh Elliot, were the discoverers of the Newfoundlands, of which there is no doubt (as now plainly appeareth) if the mariners would have been ruled then, and followed the pilot's mind..., but the lands of the West Indies, from whence all the gold cometh, had been ours, for all is one coast as by the chart appeareth."

CHAPTER 10

1.   The ancient Romans named the Atlantic after the Atlas Mountains, which folded northeast-southwest along the western end of the Mediterranean Sea and marked the limits of the known world. The Atlantic lay "beyond" the Atlas.

Atlas, in Greek mythology, was a Titan god who fought an unsuccessful war against Zeus. Zeus punished Atlas by forcing him to stand and support the sky on his shoulders forever. Atlas was portrayed standing on the northwest edge of the known world (Morocco) where the Atlas mountains are today.

2.   The contribution made by Eratosthenes in Greece 1,700 years earlier had been long forgotten.

CHAPTER 14

1.   The theory that one of Cabot's ships ran aground at Grates Cove, on the southeast coast of Newfoundland, originated partly because two Beothuk Indians were in possession of a broken sword and Venetian earring. Also, there was a rock into which Sancius Cabot, John's son, supposedly carved his name, IO Caboto and Sainmalia. In the early 1800s, it became a tourist attraction. It also attracted thousands of birds, making it a less desirable destination. In the 1960s, the rock was stolen and

disappeared in an American media van. It is now reputed to be in one of the Midwest states.

2.   In Vespucci's writings, this voyage of 1499-1500 was referred to as his "second" voyage. There is some dispute about whether he had traveled to both areas of present-day Brazil and South Carolina in 1497-1498, as some accounts claim.

3.   Looking at a map of the North Atlantic from the perspective of high latitude, it can be seen that a ship approaching North America from Great Britain would first encounter Newfoundland, and in circumnavigating the island, it would not be obvious that there was a large landmass to the west. The Grand Banks are south and east of the island, and given the apparent wealth of the fishing grounds there, the Bristol fishermen initially had no need to go any further. It appears that they credited this island as being the Island of Seven Cities.

They soon traveled further westwards, however, and apparently realized that the land to the west, which they called Brassyle, was a mainland of indeterminate size.

The French also seemed to be aware of some of these developments. A map produced in Paris in 1490, two years before Columbus sailed to the Caribbean, shows a lone island in the Atlantic Ocean that bears a strong resemblance to the shape of Newfoundland.

Chapter 16

1.   Johan Day's letter to Columbus, which lay buried in the Spanish Naval Archives until 1955, revealed that Columbus knew even before 1490 that Bristol merchants were sailing across the Atlantic to Brassyle. This letter also answered several questions that historians had been asking for several hundred years.

The Spanish language had a colloquial expression that alluded to a period of about twenty years. It was used in the letter, suggesting that Columbus knew for about "a generation" prior to 1497 when the letter was written:

> *"The said land was found and discovered 'in the past' by the men from Bristol…"*

The English translation does not adequately convey this meaning, but given that Day wrote this in 1497, it gives credence to the first voyage being in1480. More importantly, he provides the evidence that the Bristol merchants had formed a settlement or were trading in the New founde land and that Columbus knew:

> *"…as your Lordship well knows."*

He states that the Bristol merchants and Cabot saw fish drying on racks on the beaches, using exactly the same process that was developed in Iceland.

Day also mentioned that Cabot attempted a crossing in 1496 but had to turn back "because the crew was confusing him." This earlier aborted voyage had not previously been known.

CHAPTER 17

1. A brass plaque was pried loose and stolen from Johanna's grave sometime in the early 1800s. Local gossip says that it was taken by Americans, perhaps a Merrick descendant, who wanted the "crest of the daughter of America."

2.   Amerike's coat of arms consists of stars and stripes, red, white and blue. The description of the missing brass is taken from Transactions at Bristol, the Manorial History of Clifton, a nineteenth-century publication.

3.   John Broke was the son of Hugh Broke and the grandson of Sir Thomas Broke (1392-1439) of Holditch Court and Broke, Ilchester, by the heiress Cobham (Joan Baroness Cobham, Braybroke 1404-1442) in Kent. Hugh Broke owned the Canynges mansion on Redcliffe Street, the same building Amerike gave to his daughter and son-in-law. The connections between the Amerike and Broke families appear to be numerous.

The Lords Cobham were also descended from Sir Thomas Broke and included Edward Broke (1412-1464), the 6th Lord Cobham and M.P. for Somerset, and another John Broke (1464-1511), the 7th Lord Cobham.

The Cobhams have been an important family in British history both before and after the fifteenth century. Lord Cobham played a prominent role in the War of the Roses in England in the mid 1400s.

Lord Cobham fought with the Duke of York and the triumphant King Edward VI in the years between 1453 and the Battles of Mortimer's Cross and Towton in 1461. Shakespeare wrote of his involvement in these battles in "The Third Part of King Henry VI."

An earlier descendant, Lady Joan Cobham (1379-1433), also played a prominent role in history and was memorialized by Shakespeare as well.

Lady Joan Cobham was the heiress to Lord Cobham from whom the Brokes are descended. She also married Sir John Oldcastle in 1408.

Oldcastle (ca 1377-1417) became known as the good Lord Cobham and titled Baron Cobham. He was a leader of the Lollards, a dissident Christian sect that was against the power, wealth, and corruption that the church in England exhibited following the Black Death. He was born probably in Almeley, Herefordshire. In 1401, while serving in the campaign of King Henry IV to put down the Welsh rebel Owen Glendower, he became a close friend of Henry, Prince of Wales, later Henry V, King of England. Oldcastle served in the House of Commons in 1404 and in the House of Lords after 1409. Meanwhile, in defiance of royal decrees, he joined the Lollards. In 1413, the year of Prince Henry's accession to the throne, Oldcastle was convicted and condemned to death as a heretic. Henry V granted his old friend a 40-day respite in the hope that he would recant. Oldcastle escaped from imprisonment in the Tower of London and early in 1414 led an abortive revolt of the Lollards against the throne. For almost four years thereafter, he continued his Lollardian activities as a fugitive in Herefordshire. Captured on December 14, 1417, Oldcastle was executed the same day by hanging and was burned on the scaffold.

Shakespeare, following his source play, "The Famous Victories of Henry V," originally used the name Cobham in Henry IV. Probably because of protests by the current Lord Cobham, Shakespeare altered the name to Falstaff before the play was printed.

4. Sea traffic between Bristol and the American colonies, and later the United States, has been significant since 1600. Giles Penn, who came from Redcliffe, was a seafarer. His son, William, who was born in 1621, became an admiral. He was later knighted for his service to Charles II. William lent huge sums of money to King Charles, and on his death in 1670, his Quaker son, also named William, asked the king for repayment. Rather than ask for money, William asked for a grant of land in America to establish a Quaker colony. King Charles agreed as long as the land was called "Penn" in honor of his trusted admiral, hence the birth of Pennsylvania. Admiral Penn is buried at the entrance of the south transept.

Cast of Characters

Dates of birth and death are given where known.

AMERIKE, LUCY - Wife of Richard Amerike.

AMERIKE, RICHARD, ALSO AP MERYK, RICHARD, CA 1440-1503 - A wealthy landowner and merchant who traded with Spain and Portugal. He was one of the chief financial backers of the fishing expeditions to Brassyle, beginning in the 1470s, and he provided financial assistance for Cabot's expeditions to the New World on the *Matthew*. Amerike was first appointed as the King's Customs Officer in Bristol in 1486.

AP MEURIG, HYWEL, CA 1270-1330 - A Welsh prince living in the Brecon area, related to the Clanvows and bearing the same coat of arms as Richard Amerike.

BALSALL, JOHN - The resident purser on board the ship *Trinity* in the late 1470s.

BEHAIM, MARTIN 1459-1507 - A German cartographer and navigator. Behaim produced a globe in 1490 based on Toscanelli's map.

BROKE, ARTHUR - Son of Johanna and John Broke.

BROKE, DAVID 1489-1558 - Second son of Johanna and John Broke.

BROKE, HUGH CA 1515-1588 - The son of Thomas Broke and the inheritor of his estates.

BROKE, JOAN - Daughter of Johanna and John Broke.

BROKE, JOHANNA CA 1462-1538 - Daughter of Richard Amerike. Her remains are buried in St. Mary Redcliffe Church.

BROKE, JOHN CA 1460-1522 - Husband of Johanna Amerike Broke and grandson of Lord Cobham. He was a lawyer and Sergeant-at-Law to the Royal Court in London.

BROKE, THOMAS 1487-1537 - Eldest son of Johanna and John Broke.

CABOT, JOHN 1450-1499 - A celebrated navigator and European discoverer of the American mainland. Cabot gained sponsorship from the British Crown to explore a northern passage to China. He sailed on a Bristol ship called the *Matthew*.

CABOT, LUDOVICO - The son of John Cabot.

CABOT, SANCIUS CA 1480-1499 - The son of John Cabot. He accompanied his father on a voyage to the New World in 1498.

CABOT, SEBASTIAN 1476-1557 - The son of John Cabot.

CANTINO, ALBERTO - An Italian diplomat in Lisbon who illegally obtained a Portuguese map of the New World and smuggled it to his employer, the Duke of Ferrara, Ercole d'Este.

CANYNGES, THOMAS - The nephew of William Canynges. He was kidnapped by pirates and held for ransom.

CANYNGES, WILLIAM 1399-1474 - A wealthy merchant and owner of the Canynges Company, which owned the largest fleet of ships in Bristol at the time and legally controlled all fish trade with Iceland. Canynges was Mayor and a Member of Parliament for Bristol.

CLANVOW, ELIZABETH - The wife of Sir John Poyntz. They married in 1343.

COLUMBUS, BARTHOLOMEW 1461-1515 - The brother of Christopher Columbus. He and Christopher worked as chart makers and merchants together in Lisbon.

COLUMBUS, CHRISTOPHER 1451-1506 - The renowned discoverer of the Caribbean Islands, Cuba, and Santo Domingo.

COLUMBUS, DIEGO 1480-1526 - The son of Christopher Columbus.

COLUMBUS, FERNANDO 1488-1539 - The illegitimate son of Christopher Columbus.

CROFT, THOMAS 1435-1488 - The King's Customs Officer in Bristol prior to 1485 and a Member of Parliament for Leominster, Herefordshire. He was named on the trading license for the years 1480-1483. Croft also was a part-owner of the *Trinity*.

DAY, JOHAN - A Bristol merchant who wrote to Columbus about the voyages of the Bristol merchants and of John Cabot. He had lived in southern Spain and may have been working for Spanish interests.

DE ARANA, DONA BEATRIZ ENRIQUEZ - The mistress of Christopher Columbus.

DE AYALA, PEDROS - The Spanish envoy in London in 1497.

DE BALBOA, VASCO NUNEZ 1475-1517 - A Spanish conquistador and explorer. Balboa was the first European to sight the eastern shore of the Pacific Ocean.

DE LA COSA, JUAN 1450-1509 - A cartographer who explored and charted maps of the east coast of the New World. He sailed and owned the *Santa Maria* in Columbus's first expedition in 1492.

DE LA FOUNTE, WILLIAM CA 1430-1496 - A Bristol merchant who traded with Iceland as well as Spain and Portugal. He was named on the trading license for the years 1480-1483.

DE LAS CASAS, FATHER BARTHOLOMEO - The Bishop of Chiapaz in 1559 and Columbus's biographer.

DE PERANZA, DONA INES - The daughter of Christopher Columbus's mistress.

DE SANDANCOURT, JEAN BASIN - A member of the Gymnasium Vosgense.

DI SONCINO, RAIMONDO - Milan's representative in London in 1496 and a friend of John Cabot.

ELLIOT, HUGH CA 1460- - The captain of several voyages to Brassyle and the master of the ship during John Cabot's first successful voyage to the New World. He was the Sheriff of Bristol in 1501.

GUTENBERG, JOHAN 1394-1468 - The inventor of the printing press.

HOJEDA, ADMIRAL ALONZO 1470-1515 - A Spanish admiral known for his cruel and ruthless behavior. He made voyages to the New World, joining Columbus in the Caribbean. The Spanish Crown rewarded him for his patriotic work in stopping the English encroachment from the north.

ISLOND, WILLELMUS - An Icelandic national who lived in Bristol and later became an English subject. He became a merchant specializing in the Portuguese market.

JAY, HENRY - The son of John Jay. Upon his father's death, his shares in the Bristol ships passed to him and his brother, John.

JAY, JOAN CA 1425-1490 - The wife of John Jay the Elder.

JAY, JOHN CA 1395-1468 - An influential Bristol merchant who was part-owner of the *Trinity*. Jay owned several ships and had business interests in many others.

JAY, JOHN, THE ELDER CA 1420-1480 - The son of John Jay. Upon his father's death, his shares in the Bristol ships passed to him and his brother, Henry.

JAY, JOHN, III CA 1450- - The grandson of John Jay. He was on board the vessel that searched for Brassyle in 1480 and was the Sheriff of Bristol in 1499.

KEMYS, ARTHUR - Amerike's assistant at the Customs House. Cabot rented a house for his wife from Kemys.

KING EDWARD IV 1442-1483 - King of England 1461-1483.

KING EDWARD V 1470-1483 - King of England 1483.

KING HENRY VI 1421-1471 - King of England 1422-1461, 1470-1471.

KING HENRY VII 1457-1485 - King of England 1485-1509.

KING HENRY VIII 1491-1547 - King of England 1509-1547.

KING JOAO II - King of Portugal 1481-1495.

KING MANUEL I - King of Portugal 1495-1521.

KING RICHARD III 1452-1485 - King of England 1483-1485.

LLOYD - The "most skillful mariner in England" in 1480. It is believed that he skippered a voyage on the *Trinity* that took a clandestine detour to look for the island of Brassyle.

LUD, GAUTIER - The secretary to Duke Rene and a member of the Gymnasium Vosgense. Lud owned a printing press.

LUD, NICHOLAS - The nephew of Gautier Lud and a member of the Gymnasium Vosgense.

MARCHENA, FRIAR ANTONIO - A friar who lived at la Rabida Monastery in Huelva, Spain. He was an expert in astronomy, astrology, and cosmography.

MARTINEZ, CANON FERNAO - The Canon of Lisbon Cathedral in 1474.

MERCATOR, GERARD 1512-1594 - A cartographer who published a World Map in the city of Duisberg in the Duchy of Cleves in 1538. This map replaced Waldseemüller's map as the authoritative view of the earth.

PERESTRELLA, FELIPA MONEZ - The wife of Christopher Columbus. She was the aristocratic daughter of an Italian diplomat who lived in Lisbon with his Portuguese wife.

PEREZ, FRIAR JUAN - The head friar at la Rabida Monastery in Huelva, Spain, and one of Columbus's strongest advocates. Diego Columbus lived at this monastery.

POLO, MARCO 1254-1324 - A Venetian traveler and explorer. Polo was the first European to cross the entire continent of Asia and leave a record of what he saw and heard.

POYNTZ, SIR JOHN - The wealthiest and most politically powerful person in Bristol in the mid 1300s.

PTOLEMY 87-150 - An astronomer, mathematician, and geographer of the second century who considered the earth to be the center of the universe.

QUEEN ISABELLA I 1451-1501 - The Queen of Spain, 1474-1501, and wife of King Ferdinand V (1452-1516), who backed Christopher Columbus's expeditions to find a passage to India.

RENE, DUKE 1451-1508 - The Duke of Lorraine who, in order to produce a World Map, retained Waldseemüller. The Duke and the Canon of St. Die formed an intellectual group known as the Gymnasium Vosgense in 1500 to expand the rudiments of cosmography and geometry.

RINGMANN, MATTHIAS 1482-1511 - A poet and teacher of Latin and Greek and a member of the Gymnasium Vosgense.

SODERINI, PIERO 1454-   - A lifelong friend of Vespucci and the Head Magistrate of Florence during Vespucci's expeditions.

SPENCER, WILLIAM 1420-1495 - The Mayor of Bristol in 1474 and 1479 and a Member of Parliament. He was granted a trading license for the years 1480-1483.

STRAUNGE, ROBERT 1435-   - The Mayor of Bristol in 1475, 1482, and 1489 and a Member of Parliament on two occasions. He was granted a trading license.

STURMY, ROBERT - A wealthy Bristol merchant who gave his residence to the city to be used as the Cloth Hall. A portion of the building was used by the Fellowship of Merchants.

THORNE, NICHOLAS - The brother of Robert Thorne.

THORNE, NICHOLAS CA 1494-1550 - The son of Robert Thorne.

THORNE, ROBERT CA 1460-1519 - Sailed on the *Matthew* in 1497.

THORNE, ROBERT, JR. 1492-1532 - The son of Robert Thorne.

TOSCANELLI, PAOLO DEL POZZO 1397-1482 - A brilliant mathematician, astronomer, and cosmographer who produced a map showing that one could reach India by sailing west across the Atlantic Ocean.

VESPUCCI, AMERIGO 1454-1512 - An Italian discoverer who initially voyaged to the New World under the Spanish flag. Vespucci made the realization that the New World was a fourth continent. With his background in astronomy, he calculated the circumference of the earth at the equator within an accuracy of fifty miles.

WALDSEEMÜLLER, MARTIN 1470-1519 - An accomplished cosmographer and cartographer who published a revolutionary World Map in 1507.

WORCESTRE, WILLIAM CA 1415-1485 - A prolific writer of Bristol's history. He was the brother of Joan Jay.

Glossary

£-s-d. - Pounds, shillings, and pence. English money used until 1971. A pound originally was the value of a pound weight of silver. There were 20 shillings in a pound, and 12 pence in a shilling. Inflation has eroded its value by a factor of about 3,000. In the 1400s, a good wage for a skilled shipwright was 6d. a day or £7 a year. Amerike's annual salary of £6-13s -4d. would be worth approximately £20,000 or $30,000 today. Cabot's annual pension of £20 equals approximately £60,000 or $90,000. The *Trinity's* cost of £1,200 equals about £3,600,000 or $5,000,000.

ANTILLIA - Another name for the Island of Seven Cities.

AZURE - A heraldry term for blue.

BASQUES - The inhabitants of the area roughly between Bayonne in France and Bilboa in Spain. These fiercely independent people are descended from eastern Mediterranean bloodlines.

BRASSYLE, BRASYLE, BRASIL, HI-BRASSYLE, BRAZIL - An island thought to be located about 400 miles west of Ireland. It has no connection to present-day Brazil in South America.

BRAZIL WOOD - A hard wood from which a red dye was made.

BRETONS - Inhabitants of Brittany in northwest France. Brittany was once an independent country.

BROADCLOTH - Woolen cloth twenty-four yards long by two yards wide. It was used to make expensive suits.

BUSHEL - Eight gallons.

CARTOUCHE - An ornate or ornamental frame.

CATHAY - China.

CHIPANGA - Japan.

CLOTH - Woolen cloth twenty-four yards long by one yard wide.

CORRUPT WINE - Vinegar.

COSMOGRAPHY - The study of astromony, the world, and the cosmos.

COSMOLOGY - A branch of astronomy dealing with the origin and structure of the universe.

CWT - A hundred weight, 112 pounds, one-twentieth of a tun.

DENIZEN - A foreigner who is granted rights in his adopted country, similar to a working visa.

FESS - A heraldry term that means located at the mid-point of the crest.

FESTOON - An ornamental representation of a decorative chain.

GORE - A map projection that can be cut out and shaped into a globe.

GRAND BANKS - The shallow waters off the southeastern coast of Newfoundland. They were once the best fishing grounds in the world.

GULES - A heraldry term for red.

ISLAND OF SEVEN CITIES - A large island thought to be located about 1,000 miles west of Portugal.

KRILL - Minute shrimp that populate the cold northern waters and are eaten by fish.

LAST - Six-hundred and forty gallons.

MARE OCEANUM - The Atlantic Ocean. This name, given by Marco Polo, was used until the late sixteenth century.

MULLET - A heraldry term for stars on a coat of arms.

OR - A heraldry term for gold.

PALY - A heraldry term for vertical stripes on a coat of arms.

RUTTERS - Detailed sailing instructions that were handwritten in a pocket-sized book. The word comes from the French "routier" or route book. Rutters were available for most coastlines and major port-to-port destinations within Europe. They provided directions from landmark to landmark, including the bearing and distance from the previous landmark and any information relating to harbors, estuaries, depth soundings, winds, currents, sea kelp, and bird life. Fires that were maintained on headlands and hills were identified. The 100-fathom (600-foot) depth contour was also included. Rutters were given to the captain of a ship with the itinerary and sailing papers. They were used until the early sixteenth century. Printed rutters later became available.

SALT COD - Cod fish that has been filleted, compressed, and packed between layers of salt.

STOCKFISH - Air-dried cod fish that has been filleted but is joined at the tail.

STOCKS - Wooden racks from which cod fish were hung to dry.

SUBSIDY - Duty payable on imported goods.

TEREDOS - A tropical shipworm.

TRADING LICENSE - An internationally recognized document issued by a government that authorizes the holder to explore and trade within the terms of the document.

TUN OR TUNNE - A 252-gallon cask or barrel. When full of wine, it weighed approximately 2,500 lbs. Tun was also used as the unit of measurement for the carrying capacity of a ship.

WOAD - A cloth dye made from burnt wood. It was imported and used in vast quantities.

Bibliography

Andrews, K. R., WESTWARD ENTERPRISE, ENGLISH ACTIVITIES IN IRELAND, THE ATLANTIC, AND AMERICA 1480-1650, *Liverpool University Press, 1978.*

ATLAS OF WORLD HISTORY, *Harper Collins, 1999.*

Bantock, Anton, THE CABOT STORY, *Redcliffe Press, 1997.*

Bartrum, P. C., WELSH GENEOLOGIES AD 1400-1500, *National Library of Wales, 1983.*

Bedini, Silvio A., Editor, CHRISTOPHER COLUMBUS AND THE AGE OF EXPLORATION, AN ENCYCLOPEDIA, *DaCapo, 1998.*

Boorstin, Daniel J., THE DISCOVERERS, *Vintage, 1985.*

Collinson, John, HISTORY OF ANTIQUITIES OF THE COUNTY OF SOMERSET, *1741.*

Crawford, Anne, BRISTOL AND THE WINE TRADE, *Bristol Historical Association, 1984.*

Dunning, Brian, "THE MAN WHO GAVE AMERICA ITS NAME," *Country Life, June 20, 1963*

Dyson, John, and Christopher, Peter, COLUMBUS FOR GOLD, GOD, AND GLORY, *Simon and Schuster, 1991.*

Firstbrook, Peter, THE VOYAGE OF THE MATTHEW: JOHN CABOT & THE DISCOVERY OF NORTH AMERICA, *KQED, 1997.*

Fleming, Peter, and Costello, Kieran, DISCOVERING CABOT'S BRISTOL, *Redcliffe Press, 1998.*

Friar, Stephen, and Ferguson, John, BASIC HERALDRY, *Herbert, 1999.*

Godman, Colin, MA, LOWER COURT FARM ARCHAEOLOGICAL ASSESSMENT, *circa 1990.*

Hicks, Michael A., WHO'S WHO IN LATE MEDIEVAL ENGLAND, *Shepheard-Walwyn, 1991*

Jones, Donald, A HISTORY OF CLIFTON, *Phillimore Press, 1992.*

Kurlansky, Mark, COD FISHING, *Penguin Putnam, 1997.*

Macdonald, Peter, CABOT AND THE NAMING OF AMERICA, *Petmac Publications, 1997.*

Morison, Samuel Eliot, THE EUROPEAN DISCOVERY OF AMERICA: THE NORTHERN VOYAGES, *Oxford University Press, 1971.*

Morison, Samuel Eliot, THE EUROPEAN DISCOVERY OF AMERICA: THE SOUTHERN VOYAGES, *Oxford University Press, 1974.*

Nebenzahl, Kenneth, ATLAS OF COLUMBUS AND THE GREAT DISCOVERIES, *Rand McNally, 1990.*

Quinn, David, ENGLAND AND THE DISCOVERY OF AMERICA, *A. A. Knopf, 1974.*

Reddaway, T. F., and Ruddock, Alwyn A., THE ACCOUNTS OF JOHN BALSALL, PURSER OF THE TRINITY OF BRISTOL, *Royal Historical Society, 1969.*

Sansom, John, BRISTOL FIRST, *Redcliffe Press, 1997*

Sherbourne, J. W., THE PORT OF BRISTOL IN THE MIDDLE AGES, *Bristol Historical Association, 1965.*

Siddons, Michael Powell, THE DEVELOPMENT OF WELSH HERALDRY, VOL. II, A WELSH ARMORIAL, *National Library of Wales, 1993.*

Western Daily Press, "THE NAMING OF AMERICA," *August 7, 1929.*

Whitfield, Peter, THE IMAGE OF THE WORLD, *Pomegranate, 1994.*

Williamson, David, KINGS AND QUEENS OF ENGLAND, *Konecky and Konecky, 1998.*

Wilson, Ian, THE COLUMBUS MYTH, *Simon and Schuster, 1991.*

BRISTOL FILM AND VIDEO SOCIETY; http://www.bfvs.fsnet.co.uk
BRISTOL VOYAGES, Thomas, Ray; http://www.brisray.co.uk
CARTOGRAPHIC IMAGE; http://www.henry-davis.com/MAPS
THE CATHOLIC ENCYCLOPEDIA: BIOGRAPHIES OF TOSCANELLI,
BEHAIM, WALDSEEMÜLLER, VESPUCCI; http://www.newadvent.org/
cathen/
COD, Ragnow, Marguerite; http://www.bell.lib.umn.edu/Products/
cod.html
INTERNET MODERN HISTORY SOURCEBOOK;
http://modsbook.html
LOEBERTAS; http://www.loebertas.co.uk
THE MEYRICK/MERRICK COAT OF ARMS, Merrick, Robert;
http://www.knighton.freeserve.co.uk/homepage/crest.htm
THE NAMING OF AMERICA: FRAGMENTS WE'VE SHORED
AGAINST OURSELVES, Cohen, Jonathan;
http://www.uhmc.sunysb.edu/surgery/america html
NEWFOUNDLAND AND LABRADOR HERITAGE;
http://www.heritage.nf.ca
ST. MARY REDCLIFFE CHURCH; http://www.stmaryredcliffe.co.uk
TERRA INCOGNITA WEB SITE, Broome, Rodney;
http://www.ameryk.com
UNIVERSITY OF WITHYWOOD; http://www.withywood.tsx.org
WALDSEEMÜLLER MAP; http://www.bell.lib.umn.edu/map/WALD/
indexw.html
WELSH NAMES AND SURNAMES, Davies, J. B.;
http://www.korrnet.org/welsh/files/jbdavies.html

Index

A

B

G

H

King Louis XI 60
King Richard III 60
Kinsale 25, 30, 54
krill 47

L

Labrador 94
Lancaster, House of 39, 59
landmarks, naming 41–43
lateen, sail 28
Library of Congress 11
Lisbon 4, 25
Llandoger Trow 114
Lloyd, Captain 53
Longitude 96
Lorraine 2
Lud 4
    Gautier 2
    Nicholas 2

M

Madeira 27, 74
Magellan 6
Maine 88, 109
Malaga 76
Map Room 7, 41, 58, 66
Marchena, Friar 26
Mare Oceanum 3, 27, 54, 63, 65, 80
Mars 96

P

Q

R

V

W

Y